12

Algebraic Methods for Signal Processing and Communications Coding

Signal Processing and Digital Filtering

Synthethic Aperture Radar
J.P. Fitch

Multiplicative Complexity, Convolution and the DFT
M.T. Heideman

Array Signal Processing
S.U. Pillai

Maximum Likelihood Deconvolution
J.M. Mendel

Algorithms for Discrete Fourier Transform and Convolution
R. Tolimieri, M. An, and C. Lu

Algebraic Methods for Signal Processing and Communications Coding
R.E. Blahut

Richard E. Blahut

Algebraic Methods for Signal Processing and Communications Coding

C.S. Burrus
Consulting Editor

With 12 Illustrations

Springer-Verlag
New York Berlin Heidelberg London Paris
Tokyo Hong Kong Barcelona Budapest

Richard E. Blahut
IBM
Owego, NY
USA

Consulting Editor
Signal Processing and Digital Filtering

C.S. Burrus
Professor and Chairman
Department of Electrical
        and Computer Engineering
Rice University
Houston, TX 77251-1892
USA

Library of Congress Cataloging-in-Publication Data
Blahut, Richard E.
  Algebraic methods for signal processing and communications coding
/ Richard E. Blahut.
    p. cm.
  Includes bibliographical references (p.   ) and index.
  ISBN 0-387-97673-6 (Springer-Verlag New York Berlin Heidelberg :
acid-free paper). -- ISBN 3-540-97673-6 (Springer-Verlag Berlin
Heidelberg New York : acid-free paper)
    1. Signal processing--Digital techniques. 2 . Coding theory.
  I. Title.
  TK5102.5.B534 1992
  621.382'2--dc20   91-37274

Printed on acid-free paper.

Production managed by Karen Phillips; manufacturing supervised by Robert Paella.
Photocomposed pages prepared using L^{A}T$_E$X.
Printed and bound by Braun-Brumfield, Ann Arbor, MI.
Printed in the United States of America.

9 8 7 6 5 4 3 2 1

ISBN 0-387-97673-6 Springer-Verlag New York Berlin Heidelberg
ISBN 3-540-97673-6 Springer-Verlag Berlin Heidelberg New York

If I had more time I could write a shorter letter.

- *Blaise Pascal*

Preface

Algorithms for computation are a central part of both digital signal processing and decoders for error-control codes and the central algorithms of the two subjects share many similarities. Each subject makes extensive use of the discrete Fourier transform, of convolutions, and of algorithms for the inversion of Toeplitz systems of equations.

Digital signal processing is now an established subject in its own right; it no longer needs to be viewed as a digitized version of analog signal processing. Algebraic structures are becoming more important to its development. Many of the techniques of digital signal processing are valid in any algebraic field, although in most cases at least part of the problem will naturally lie either in the real field or the complex field because that is where the data originate. In other cases the choice of field for computations may be up to the algorithm designer, who usually chooses the real field or the complex field because of familiarity with it or because it is suitable for the particular application. Still, it is appropriate to catalog the many algebraic fields in a way that is accessible to students of digital signal processing, in hopes of stimulating new applications to engineering tasks.

An error-control code can be viewed as a technique of digital signal processing intended to prevent random errors or burst errors caused by noise in a communication channel. Error-control codes can be developed in an arbitrary field; this is so despite the fact that finite fields of characteristic 2, $GF(2^m)$, seem to be the best choice in most applications. Even though the underlying data may originate as real numbers and the codewords must be represented as another sequence of real or complex numbers for modulation into a communication waveform, it is better to represent the data temporarily by elements of a finite field for coding. However, there has not been much study of error-control codes in algebraic fields other than finite fields. Further insights should emerge as time goes by.

The primary purpose of this monograph is to explore the ties between digital signal processing and error-control codes, with the thought of eventually making them the two components of a unified theory, or of making a large part of the theory of error-control codes a subset of digital signal processing. By studying the properties of the Fourier transform in an arbitrary field, a perspective emerges in which the two subjects are unified. Because there are many fields and many Fourier transforms in most of those fields, the unified view will also uncover a rich set of mathematical tools, many of which have yet to find an engineering application.

The secondary purposes of the monograph are of four kinds: First, a parallel development of the two subjects makes the subject of error-control coding more accessible to the many engineers familiar with digital signal processing. Second, the many fast algorithms of digital signal processing can be borrowed for use in the field of error-control codes. Third, the techniques of error-control can be borrowed for use in the field of digital signal processing to protect against impulsive noise. And fourth, perhaps the design of VLSI chips that perform a variety of signal processing and error-control tasks would become simplified if these tasks could be merged and performed in the same algebraic field.

The monograph attempts to combine many elements from digital signal processing and error-control codes into a single study. I hope that this attempt at unification justifies repeating some of these topics here. In part the book is based on a short course I have given in Zurich under the sponsorship of Advanced Technology Seminars in which aspects of digital signal processing were integrated with topics on error-control codes. I am grateful to Lis Massey for organizing that course, to Ephraim Feig for providing comments on a draft of this monograph, to Richard Games for providing illustrations of the algebraic integers embedded in the complex plane, and to IBM for allowing me the time to write this monograph.

Contents

1

Introduction

The processing of a digital communication system uses distinct forms of digital signal processing such as a digital filter and an error-control code. To what extent do digital signal processing and error-control coding really differ? Digital signal processing is a subject whose usual setting is the real or the complex number system, denoted **R** or **C**, while error-control coding is a subject whose usual setting is a different number system called a Galois field, denoted $GF(q)$. The number systems denoted **R**, **C**, and $GF(q)$ are examples of the general mathematical structure called an *algebraic field* which can be used for a variety of computational and engineering purposes. Our goal is to guide a diverse audience of engineers through the topics of digital signal processing and error-control codes in a variety of algebraic fields, in an effort to partly unify the techniques, to expose new computational methods, and to open new applications. Both subjects are strongly dependent on the discrete Fourier transform in their respective algebraic fields.

The computational tasks at the center of applied digital signal processing are convolutions, Fourier transforms, and the inversion of Toeplitz systems of equations for spectral estimation. Similarly, if the language is appropriately chosen, the computational tasks at the center of applied error-control coding are also convolutions, Fourier transforms, and the inversion of Toeplitz systems of equations, albeit in a different field — in this case a Galois field. Thus from the computational point of view, we should expect valuable insights from laying the two subjects side by side.

There are also some interesting parallels in the goals of the two subjects. Much of digital signal processing is devoted to the task of removing noise and interference, usually nonimpulsive, by passing a known signal through a suitable filter. Error-control coding, especially if it is reformulated in the real or complex field, can be described as the task of removing impulsive or burst noise from a specially designed signal by a nonlinear filtering operation. By pushing this view, we intend to blur the demarcation between digital signal processing and error control, though we stop short of saying that the two subjects are really the same.

The central threads running through the book are the discrete Fourier transform and the Toeplitz matrix. Most properties of the Fourier transform are valid in many number systems, particularly those called algebraic fields. Similarly, most methods for the inversion of a Toeplitz system of equations are valid in any algebraic field. Study of the general case affords us the broader and more unified view. We will emphasize especially how the general properties of the discrete Fourier transform do not depend on the field. In particular, transforms that go by other names, such as number theory transforms and polynomial transforms, are actually Fourier transforms but defined in a different algebraic field. By emphasizing the general notion of a field, many transforms are unified.

1.1   Digital Signal Processing

Digital signal processing is now becoming a mature subject. Its major tasks are to process a block of data or an unbounded stream of data to suppress noise or interference, to estimate spectra or images, and to estimate parameters of interest from the data stream. Common computations of digital signal processing include computing the output of a finite impulse response (FIR) filter and computing the coefficients of a correlation. A FIR filter is described by the acyclic convolution

$$s_i = \sum_{k=0}^{L-1} g_k d_{i-k},$$

where $\dots, d_{-1}, d_0, d_1, \dots$ is the *input data sequence*, $\dots, s_{-1}, s_0, s_1, \dots$ is the *output data sequence* and $g_0, g_1, \dots, g_{L-1}$ are the *weights* of the L *taps* of the FIR filter.

The acyclic convolution of the FIR filter resembles the cyclic convolution

$$s_i = \sum_{k=0}^{n-1} g_k d_{((i-k))},$$

where the double parentheses denote modulo-n indexing and the vectors **d**, **g**, and **s** are of length n. Although the appearance is similar, the properties of the acyclic convolution and the cyclic convolution can be quite different.

Normally, in either convolution, the data and the tap weights consist of fixed-point numbers with b bits of wordlength. In fact, by temporarily rescaling the b bits we can place the decimal point at the right. Thus these computational tasks usually can be regarded as merely the convolution of sequences of integer data to obtain sequences of integer data. Nevertheless, it is considered natural to embed the integer computations into the complex field so that properties of the discrete Fourier transform can be exploited to reduce the computational load.

An important consideration is that of computational precision. While integer convolution is, in principle, exact — even in a finite-state machine — the introduction of certain popular fast convolution algorithms in the complex field brings on problems of roundoff, and precision errors may ensue.

One can also embed the problem of integer convolution into the field of modulo-p integer arithmetic, denoted $GF(p)$, and use a "number theory transform," but some consider this to be a somewhat unnatural expedient partly because the "true" arithmetic interpretation of the data is suppressed during the computation.

An important consideration when using $GF(p)$ to compute an integer convolution is wordlength. In finite fields there is no natural mechanism for discarding low-order digits at intermediate points of the computation. Therefore, while the numbers are considered to lie in $GF(p)$, larger wordlength may be necessary than the final integer answer requires.

A second computational task of digital signal processing is the computation of the discrete Fourier transform. A Fourier transform can be defined in any field. Let $\mathbf{v}$ be a vector of blocklength n over the field F and let ω be an element of order n in F (or an extension of F). The Fourier transform is given by the expression

$$V_k = \sum_{i=0}^{n-1} \omega^{ik} v_i, \qquad k = 0, \ldots, n-1.$$

The Fourier transform produces the new vector $\mathbf{V}$, also of blocklength n, in the field F (or an extension of F).

While the expression for the discrete Fourier transform is quite simple, there is a surprisingly extensive literature discussing its computation and applications. The Fourier transform can be used either for spectral analysis or as an indirect method of computing convolutions. Sometimes, to compute a convolution in one field, the variables may be moved into a surrogate field for the computation. Thus a Fourier transform in one algebraic field may be used to compute a convolution in a different field.

The third task of digital signal processing that we shall study is the inversion of Toeplitz systems of equations. A Toeplitz matrix (over any algebraic field F) is a matrix of the form

$$\mathbf{A} = \begin{bmatrix} a_0 & a_1 & a_2 & \cdots & a_{n-1} \\ a_{-1} & a_0 & a_1 & \cdots & a_{n-2} \\ a_{-2} & a_{-1} & a_0 & \cdots & a_{n-3} \\ \vdots & & & & \vdots \\ a_{-n+1} & & & \cdots & a_0 \end{bmatrix}$$

with equal elements running down each minor diagonal. The a_{ij} element can be expressed as

$$a_{ij} = a_{i-j}.$$

Only $2n-1$ elements of the field F are needed to specify an $n \times n$ Toeplitz matrix over F.

A Toeplitz system of equations is a matrix-vector equation of the form

$$\mathbf{A}\mathbf{f} = \mathbf{g},$$

where $\mathbf{A}$ is an $n \times n$ Toeplitz matrix, and $\mathbf{f}$ and $\mathbf{g}$ are n vectors. To solve a Toeplitz system of equations, one must find $\mathbf{f}$ when given $\mathbf{A}$ and $\mathbf{g}$.

Fast algorithms for the inversion of Toeplitz systems of equations are highly valued by those who use them. The algorithms of Levinson and of Durbin are well known within the subject of digital signal processing. The algorithms of Berlekamp and Massey and of Sugiyama are well known within the literature of error-control codes. These latter algorithms have been most widely used in finite fields, but they are valid in any field.

1.2   Digital Communications

The primary task of digital communication theory is the design of waveforms for the transmission of data from one place to another. A *data transmission code* is a procedure for forming a message as a sequence of symbols from a discrete alphabet that is resistant to errors or noise occurring in the communication channel. A data transmission code can be described as a technique of digital signal processing that will protect a discrete-time waveform from impulsive noise or burst noise. A data transmission code can have a block structure or a tree structure. A data transmission code is also called an *error-control code* if it is designed to resist explicit symbol errors as opposed to general background noise.

The simplest block code for error control is the binary code known as the Hamming (7,4) code and shown in Figure 1.1. This code has 16 binary seven-bit codewords corresponding to four data bits. To form a codeword from the data bits, which we may take to be the first four bits of each codeword, three additional bits called parity-check bits are appended to the data bits.

In the general block code, the symbols are usually from an algebraic number system known as a Galois field or a finite field and denoted by the label $GF(q)$. To every k data symbols taken from the field $GF(q)$, a codeword is formed by appending $n - k$ additional symbols called parity-check symbols. The codeword now has blocklength n and is constructed so that the original k data symbols can be recovered even when up to t symbols are corrupted by additive channel noise. This is called a t-error-correcting block code. The relationship between t, k, and n for some codes of interest will be developed in due course.

In contrast to the computations of digital signal processing, those of most error-control codes take place in the finite field $GF(q)$. This setting is considered natural partly because in error-control problems the data consist merely of m-bit binary words without any intrinsic notion of algebraic structure. We are free to embed these data words into any convenient algebraic field to get enough arithmetic structure so that codes can be defined and the encoding and decoding calculations can be performed.

A communication code can be defined in any of the algebraic fields that

```
0  0  0  0  0  0  0
0  0  0  1  0  1  1
0  0  1  0  1  1  0
0  0  1  1  1  0  1
0  1  0  0  1  1  1
0  1  0  1  1  0  0
0  1  1  0  0  0  1
0  1  1  1  0  1  0
1  0  0  0  1  0  1
1  0  0  1  1  1  0
1  0  1  0  0  1  1
1  0  1  1  0  0  0
1  1  0  0  0  1  0
1  1  0  1  0  0  1
1  1  1  0  1  0  0
1  1  1  1  1  1  1
```

FIGURE 1.1. Hamming (7,4) code.

will be studied in Chapter 2. Many engineering issues may influence the choice of field. Many data sources are intrinsically binary data sources. The data symbols can be represented neatly in a finite field $GF(2^m)$, but require some form of modulation and demodulation to match them to the real or complex numbers used by channels. Many channels are intrinsically real or complex channels and so require some form of modulation and demodulation to match them to a finite field. For example, a BPSK or QPSK modulator and demodulator may be used to convert the real-valued or complex-valued inputs and outputs of a communication channel into binary or quaternary symbols suitable for a code in a finite field.

2

Mathematical Fundamentals

Algebraic structures and number theory underlie much of the development of modern techniques of digital signal processing and digital communications, so it behooves us to begin our study with a formal survey of these mathematical fundamentals. Although the familiar real and complex number systems satisfy most of our everyday needs, we will find practical uses for other number systems. Therefore we must develop the theories of these alternative number systems. Principally, we shall study a *field* in which addition, subtraction, multiplication, and division are defined, a *ring* in which addition, subtraction, and multiplication are defined, and a *group* in which a single operation and its inverse are defined.

2.1 Algebraic Fields

Loosely, a field is any arithmetic system in which one can add, subtract, multiply, and divide such that the usual arithmetic properties of associativity, distributivity, and commutativity are satisfied. In algebraic fields that we are familiar with, the operations of addition, subtraction, multiplication, and division have become routine. Most readers will be familiar with the rational field — denoted $\mathbf{Q}$ — and consisting of all numbers of the form a/b, where a and b are integers; with the real field — denoted $\mathbf{R}$ — and consisting of all finite or infinite decimals, and with the complex field — denoted $\mathbf{C}$ — and consisting of all numbers of the form $a + jb$ where a and b are real. In each of these fields the addition and multiplication rules, as well as the rules of subtraction and division, are well known.

Less well known is the field of complex rationals, denoted $\mathbf{Q}^2$ or $\mathbf{Q}(j)$. This field consists of all numbers of the form $a + jb$, where a and b are rational; addition and multiplication are as complex numbers. It is easy to see that $\mathbf{Q}^2$ is closed under addition and multiplication as well as under their inverses, subtraction and division.

Each of these fields contains an infinite number of numbers. It is common to study the theory of digital signal processing in the real or complex field even though in applications, problems of digital signal processing involve only a finite set of numbers — perhaps the set of 16-bit fixed-point numbers. This set of 16-bit numbers under the conventional definitions of addition and multiplication does not form a field because the system is not closed. Roundoff and overflow always occur in conventional 16-bit arith-metic. One may wish to define arithmetic operations to make a finite set of numbers into a field and thereby eliminate roundoff and overflow. This is often possible but only with seemingly bizarre definitions of addition and multiplication. To eliminate roundoff and overflow the arithmetic system must give up its everyday character and much of its practical usefulness. Despite these problems such fields are often useful for many tasks.

In algebraic fields that we may be unfamiliar with, the operations of ad-dition, subtraction, multiplication, and division will appear strange. How-ever, whenever we are successful in defining a field, the familiar structures of algebra will be valid because they depend on the formal properties of distributivity, associativity, and commutativity but not on the specific def-initions of the arithmetic operations.

A formal definition of a field is as follows. A field F is a set that has two arithmetic operations defined on it: addition and multiplication, such that the following properties are satisfied.

1) (Addition Axiom) The set F is closed under addition, and addition is associative and commutative

$$
\begin{aligned}
a + (b + c) &= (a + b) + c, \\
a + b &= b + a.
\end{aligned}
$$

There is an element called *zero* and denoted 0 such that $a + 0 = a$, and every element a has an element called the *negative* of a and denoted $(-a)$ such that $a + (-a) = 0$. Subtraction $a - b$ is defined as $a + (-b)$.

2) (Multiplication Axiom) The set is closed under multiplication, and multiplication is associative and commutative

$$
\begin{aligned}
a(bc) &= (ab)c, \\
ab &= ba.
\end{aligned}
$$

There is an element called *one* and denoted 1 such that $1a = a$ and for every element a except zero there is an element called the *inverse* of a and denoted a^{-1} such that $aa^{-1} = 1$. Division $a \div b$ (or a/b) is defined as ab^{-1}.

3) (Joint Axiom) The distributive law

$$
(a + b)c = ac + bc
$$

holds for all numbers a, b, and c in the set F.

In addition to those fields with an infinite number of elements, there are also fields with a finite number of elements; we shall find uses for these as well. A field with q elements, if there is one, is called a *finite field* or a *Galois field* and is denoted by the label $GF(q)$. The field with q elements is unique (except for notation), so the label $GF(q)$ suffices.

Every field must have an element 0 and an element 1, so that every field has at least two elements. In fact with the addition and multiplication tables,

+	0	1
0	0	1
1	1	0

·	0	1
0	0	0
1	0	1

these suffice to form a field. This is the field known as $GF(2)$.

Whenever p is a prime, we can construct the finite field $GF(p)$ as the set $GF(p) = \{0, 1, 2, \ldots, p-1\}$ together with modulo-p addition and modulo-p multiplication. This structure forms a field. No other field with p elements exists. More formally, any other field with p elements is the same as this one except that different names may be assigned to the elements. The technical way of saying this is that all fields with p elements are *isomorphic*.

When p is small enough, we can explicitly write out the addition and multiplication tables. For example, the field $GF(3)$ consists of the elements $\{0, 1, 2\}$ together with the operations

+	0	1	2
0	0	1	2
1	1	2	0
2	2	0	1

·	0	1	2
0	0	0	0
1	0	1	2
2	0	2	1

The field $GF(5)$ consists of the elements $\{0, 1, 2, 3, 4\}$ with the operations

+	0	1	2	3	4
0	0	1	2	3	4
1	1	2	3	4	0
2	2	3	4	0	1
3	3	4	0	1	2
4	4	0	1	2	3

·	0	1	2	3	4
0	0	0	0	0	0
1	0	1	2	3	4
2	0	2	4	1	3
3	0	3	1	4	2
4	0	4	3	2	1

These examples are very small fields. We will find uses for much larger prime fields such as $GF(2^{16} + 1)$ and $GF(2^{13} - 1)$ that can hold 16-bit or 12-bit integers, as well as some fields $GF(q)$ in which the number of elements q is not a prime. This latter case is not based on modulo-q arithmetic; we shall describe its structure in due course. In general, $GF(q)$ is not based on modulo-q arithmetic whenever q is not a prime. Furthermore, $GF(q)$ exists only if q is a prime or a power of a prime.

Every finite field $GF(q)$ has the important, though difficult to prove, property that the set of nonzero field elements can be expressed as $q - 1$ powers of a fixed element π, called a *primitive element* of $GF(q)$. Thus we can write

$$GF(q) = \{0, \pi^0, \pi^1, \ldots, \pi^{q-2}\}.$$

For example,

$$GF(5) = \{0, 1, 2, 3, 4\}$$
$$= \{0, 2^0, 2^1, 2^2, 2^3\}.$$

When field elements are expressed as powers of π, multiplication consists of adding exponents. Thus, if $\alpha = \pi^i$ and $\beta = \pi^j$, then $\alpha\beta = \pi^{i+j}$.

In any field, most of the familiar techniques of linear algebra can be used. Matrix theory, including the definition and computation of determinants and matrix inverses, is valid in any field. This will be easy to accept if we think back to the development of these matrix operations. Their development relied on algebraic equations involving addition and multiplication. The validity of the development depends only on the formal properties of the operations, but not on what addition and multiplication "really" are.

An algebraic field can be classified by an integer called its characteristic which is defined as follows. Every field has an element 1 which can be added to itself any number of times to get the sequence

$$1, 1 + 1, 1 + 1 + 1, 1 + 1 + 1 + 1, \ldots .$$

Either the sequence continues to infinity without repetition, or at some point the first repetition occurs. The element before the first repetition must be the element 0 (otherwise it would itself be a repetition). That is,

$$1 + 1 + \ldots + 1 = 0,$$

where there are a finite number of 1's in the sum. The smallest number of 1's that sum to zero in the field F is called the *characteristic* of the field F. (The characteristic of $GF(5)$ is five.) If there is no such number the field characteristic is defined to be infinity (or by some people, zero). Whenever the field characteristic is finite, the characteristic must be a prime number.

It will help to make algebraic fields more understandable if we pursue this construction a little further. If the field characteristic is infinite then the sequence of elements

$$1, 1 + 1, 1 + 1 + 1, 1 + 1 + 1 + 1, \ldots$$

never repeats. We may name the terms with the notation

$$1, 2, 3, 4, \ldots$$

and call them the *integers* of F. Because F is a field, the ratio of any two integers must also be in F. Therefore, F contains a subset of elements that behaves very much like the field of rationals. This is expressed formally by the statement that every field of infinite characteristic contains a subfield isomorphic to the rational field.

If the field has characteristic p, then we have the sequence

$$1, 1 + 1, 1 + 1 + 1, \ldots, 0.$$

We may name these elements with the notation

$$1, 2, 3, \ldots, p - 1, 0,$$

which again are called the *integers* of the field F. If we continue the sequence, it repeats with period p, which means that addition of two integers is modulo-p addition. This, in turn, means that multiplication of two integers is modulo-p multiplication because the product of integers ab can be written

$$\begin{aligned} ab &= (1 + 1 + \ldots + 1)b \\ &= b + b + \ldots + b, \end{aligned}$$

where there are a copies of b on the right.

The set of integers of a field of finite characteristic is in fact itself a field because modulo-p arithmetic does form a field denoted $GF(p)$ whenever p is a prime. The formal statement is that every field of characteristic p, p finite, contains a subfield isomorphic to $GF(p)$.

The Galois fields $GF(q)$ in which the operation of multiplication is simple include those $GF(p)$ for which p is a prime such that the modulo-p operation is simple to compute. When implemented with binary logic, the modulo-p operation is simple if the prime p is of the form $2^m \pm 1$ because $2^m \pmod{2^m + 1} = -1$ and $2^m \pmod{2^m - 1} = +1$.

The field $GF(2^m + 1)$ exists whenever $p = 2^m + 1$ is a prime. If m is not a binary power, then $2^m + 1$ is not a prime. If m is a binary power, then $2^m + 1$ may be a prime, but not always; $2^{32} + 1$ is known to be composite. However, primes are found for $m = 2$, 4, 8, or 16 for which $2^m + 1 = 5$, 17, 257, or 65,537; these integers are known as *Fermat primes*. Arithmetic in the Fermat field $GF(2^m + 1)$ is convenient if the integers are represented as m-bit binary numbers. Because in this field $2^m + 1 = 0$, the overflow 2^m is equal to -1. Hence, the arithmetic is conventional integer arithmetic and the overflow bits are subtracted from the low-order bits of the number.

The field $GF(2^m - 1)$ exists whenever $p = 2^m - 1$ is a prime. This requires that m be a prime; because $2^{ab} - 1$ is divisible by $2^a - 1$, $2^m - 1$ cannot be a prime if m is composite. The smallest values of m for which $2^m - 1$ is a prime are 3, 5, 7, 13, 17, 19, and 31, for which $2^m - 1 = 7$, 31, 127, 8191, 131,071, 524,287, and 2,147,483,647; these integers are known as *Mersenne primes*. Notice that $2^{11} - 1$ is not a prime although 11 is a prime, so a field is not guaranteed when m is a prime. Arithmetic in the Mersenne field $GF(2^m - 1)$ is convenient if the integers are represented as m-bit binary numbers. Because in this field, $2^m - 1 = 0$, the overflow 2^m is equal to 1. Hence, the arithmetic is conventional integer arithmetic and the overflow bits are added into the low-order bits of the number.

2.2 Elementary Number Theory

In the set of integers, which is denoted $\mathbf{Z}$, division is not defined in general. The integer b is said to *divide* the integer c or be a *factor* of c if there exists an integer a such that $c = ab$. The *division algorithm* says that given any integer c and any positive integer d, there exist unique integers Q, called the *quotient*, and s, called the *remainder*, such that $c = dQ + s$ and $0 \leq s < d$. The quotient and remainder may be denoted as

$$Q = \left\lfloor \frac{c}{d} \right\rfloor, \qquad s = R_d[c].$$

The *greatest common divisor* of two integers s and t, denoted $\text{GCD}(s,t)$, is the largest integer that divides both s and t. Two integers s and t are called *relatively prime* if $\text{GCD}(s,t) = 1$. A positive integer p is *prime* if it is relatively prime with every smaller positive integer. An integer is *composite* if it is not prime.

The *Euclidean algorithm* is a recursive procedure to find the greatest common divisor of two integers s and $t < s$. Let $s^{(0)} = s$ and $t^{(0)} = t$. The recursion is

$$\begin{bmatrix} s^{(r)} \\ t^{(r)} \end{bmatrix} = \begin{bmatrix} 0 & 1 \\ 1 & -Q^{(r)} \end{bmatrix} \begin{bmatrix} s^{(r-1)} \\ t^{(r-1)} \end{bmatrix}$$

as long as $t^{(r-1)}$ is nonzero, and where

$$Q^{(r)} = \left\lfloor \frac{s^{(r-1)}}{t^{(r-1)}} \right\rfloor$$

is the quotient obtained when $s^{(r-1)}$ is divided by $t^{(r-1)}$. When $t^{(r-1)}$ is zero, $\text{GCD}(s,t) = s^{(r-1)}$.

The 2×2 matrix of the recursion has determinant equal to -1. Therefore the inverse of that matrix has only integer elements as does a product of such inverse matrices. Therefore

$$\begin{bmatrix} s^{(r-1)} \\ t^{(r-1)} \end{bmatrix} = \mathbf{A} \begin{bmatrix} s \\ t \end{bmatrix},$$

where $\mathbf{A}$ is a 2×2 matrix with integer elements. Consequently,

$$\text{GCD}(s,t) = as + bt$$

for some integers a and b. In particular, if s and t are relatively prime, the equation

$$as + bt = 1$$

is satisfied by some integers a and b.

The *Chinese remainder theorem* for integers says that for any set of pairwise relatively prime, positive integers $\{m_0, m_1, \ldots, m_{K-1}\}$ the set of congruences

$$c \equiv c_k \pmod{m_k}, \qquad k = 0, \ldots, K-1$$

has exactly one nonnegative solution smaller than the product $\prod_k m_k$. The set of congruences can be inverted by the equation

$$c = \sum_{k=0}^{K-1} c_k N_k M_k \pmod{M},$$

where $M = \prod_k m_k$, $M_k = M/m_k$, and N_k solves

$$N_k M_k + n_k m_k = 1.$$

Such an N_k must exist because $\text{GCD}(M_k, m_k) = 1$.

It will sometimes be necessary to state how many integers smaller than q are relatively prime to q. The *totient function*, denoted $\phi(q)$, is defined (when q is larger than 1) as the number of nonzero integers smaller than q that are relatively prime to q. For $q = 1$, $\phi(q) = 1$.

When q is a prime p, then all the nonzero integers smaller than p are relatively prime to p, and so $\phi(p) = p - 1$ whenever p is a prime. When q is a power of a prime p^m, then the only integers smaller than p^m not relatively prime to p^m are the p^{m-1} multiples of p. Therefore

$$\phi(p^m) = p^m - p^{m-1} = p^{m-1}(p - 1).$$

Other values of the totient function are described by the following theorem.

☐ **Theorem 2.2.1** *If* $\text{GCD}(q', q'') = 1$, *then*

$$\phi(q'q'') = \phi(q')\phi(q'').$$

Proof The nonnegative integers smaller than $q'q''$ are $i = 0, \ldots, q'q'' - 1$. Imagine these integers mapped into a two-dimensional array using indices given by

$$i' = i \pmod{q'},$$
$$i'' = i \pmod{q''}.$$

Because $\text{GCD}(q', q'') = 1$, the mapping from i to pairs (i', i'') is one to one. For some Q' and Q'',

$$i = q'Q' + i',$$
$$i = q''Q'' + i''.$$

Suppose that i' has no factor in common with q' and i'' has no factor in common with q''; then i has no factor in common with either q' or q'' and hence no factor in common with $q'q''$. Therefore

$$\phi(q'q'') \geq \phi(q')\phi(q'').$$

Conversely, if i has no factor in common with $q'q''$ then it has no factor in common with q' or with q''. Hence i' has no factor in common with q', and i'' has no factor in common with q''. That is,

$$\phi(q'q'') \le \phi(q')\phi(q'').$$

and the theorem is proved. □

□ **Corollary 2.2.2** If $q = p_1^{c_1} p_2^{c_2} \ldots p_r^{c_r}$ is the prime factorization of q, then

$$\phi(q) = p_1^{c_1-1} p_2^{c_2-1} \ldots p_r^{c_r-1}(p_1 - 1)(p_2 - 1) \ldots (p_r - 1). \qquad □$$

There is another important relationship satisfied by the totient function that will prove useful. Suppose that d is a divisor of q, and suppose that $f(x)$ is any function. The sum of all terms $f(d)$ such that d is a divisor of q is written $\sum_{d|q} f(d)$. It is clear that

$$\sum_{d|q} \phi(d) = \sum_{d|q} \phi\left(\frac{q}{d}\right)$$

because $\left(\frac{q}{d}\right)$ divides q whenever d divides q.

□ **Theorem 2.2.3** The totient function satisfies

$$\sum_{d|q} \phi(d) = q.$$

Proof For each d that divides q, consider the following set of nonnegative integers smaller than q:

$$\{i \mid \mathrm{GCD}(i, q) = d\}.$$

Each integer smaller than q will belong to exactly one such set. Hence if we sum the number of elements in each set, we get q.

Now consider the equivalent definition of the set

$$\left\{i \mid \mathrm{GCD}\left(\frac{i}{d}, \frac{q}{d}\right) = 1\right\}.$$

This set has exactly $\phi\left(\frac{q}{d}\right)$ elements. Summing gives

$$\sum_{d|q} \phi\left(\frac{q}{d}\right) = q$$

and the theorem follows. □

In the prime field $GF(p)$, the element -1 is a square if $p = 1 \pmod 4$, and is not a square if $p = -1 \pmod 4$. To prove this let π be a primitive element of $GF(p)$. Then the $p-1$ nonzero elements of $GF(p)$ can be written as the sequence

$$\pi^1, \pi^2, \pi^3, \ldots, -1, \ldots, \pi^{p-2}, 1$$

in which -1 appears somewhere in the sequence and nonsquares and squares alternate. The latter condition is true because only half the elements can be squares (since β and $-\beta$ have the same square) and all even powers of π are clearly squares. Because $\pi^{p-1} = 1$ and $(-1)^2 = 1$, it is clear that

$$\pi^{(p-1)/2} = -1.$$

If $(p-1)/2$ is even, -1 appears in the above sequence in the position of a square, otherwise it appears in the position of a nonsquare.

Let p be a fixed odd prime. The *Legendre symbol*, denoted $\chi(i)$, is defined by

$$\chi(i) = \begin{cases} 0 & \text{if } i \text{ equals zero} \quad (\text{mod } p), \\ 1 & \text{if } i \text{ is a nonzero square} \quad (\text{mod } p), \\ -1 & \text{if } i \text{ is a nonzero nonsquare} \quad (\text{mod } p). \end{cases}$$

An important property of the Legendre symbol that we shall use is $\chi(ij) = \chi(i)\chi(j)$. To prove this let $i = \pi^r$ and $j = \pi^s$ where π is a primitive element of $GF(p)$. Then $ij = \pi^{r+s}$, which is a square if $r + s$ is even.

Let p be a fixed odd prime. Let F be any field and ω an element of F (or an extension of F) of order p. Such an ω exists provided the characteristic of F is not p. The *Gaussian sum*, denoted θ, is defined as

$$\theta = \sum_{i=0}^{p-1} \chi(i)\omega^i.$$

It is the first component of the Fourier transform of the vector with components of $\chi(i)$. We shall see in the next chapter that regardless of the field F, $\theta^2 = -p$ if $p \equiv -1 \pmod 4$ and $\theta^2 = p$ if $p \equiv 1 \pmod 4$.

2.3   Extension Fields

Most fields can be extended to create a larger field. The new field then is called an *extension field* of the original field. An orderly way of extending fields is by using a construction called the polynomial representation of the extension field. To see how this works, we extend the real field to get the complex field. The method is to find a polynomial of degree 2 over the real field that cannot be factored over the real field. As long as we are restricted to the real field, the polynomial

$$p(x) = x^2 + 1$$

is irreducible — it cannot be factored, nor does it have any zeros. Not to be discouraged, we invent a zero called an imaginary number and denoted by the symbol j.[1] Then we have the factorization

$$p(x) = (x + j)(x - j)$$

if we simply declare that $j^2 = -1$. Now we have a new number system with elements of the form $(a + jb)$ and it turns out that this is a field because all nonzero elements have inverses given by

$$(a + jb)^{-1} = \frac{a}{a^2 + b^2} - j\frac{b}{a^2 + b^2},$$

where the operations within each component are operations of the real number system.

To introduce the way to develop other extension fields, we shall represent the elements of the complex field instead as objects in the form of polynomials $a + bx$, where a and b are real numbers. This merely requires that the symbol x be used in place of the symbol j. Multiplication of these field elements now is a polynomial product modulo the polynomial $x^2 + 1$. That is, the product is of the form

$$e + fx = (a + bx)(c + dx) \qquad (\text{mod } x^2 + 1).$$

The operation modulo the polynomial $x^2 + 1$ means to divide by $x^2 + 1$ and keep the remainder. But this operation turns the x^2 term of the product into -1 so the polynomial product modulo the polynomial $x^2 + 1$ is the same structure as complex multiplication.

The general construction for extending a field F works as follows. Instead of using just two components, as the real and imaginary parts of the above example, there can be m components. Thus the elements of the extension field are the set of all m-tuples from the field F, where m is a fixed integer. We denote this set by F^m. Thus

$$F^m = \{(a_{m-1}, a_{m-2}, \ldots, a_1, a_0)\},$$

where each component of the m-tuple is an element of the field F. We will represent the elements of F^m more usefully in the form of polynomials:

$$F^m = \{(a_{m-1}x^{m-1} + a_{m-2}x^{m-2} + \ldots + a_1x + a_0)\}.$$

The symbol x can be thought of simply as a place marker. To make this set into a field we need to define addition and multiplication. Addition is defined as addition of polynomials (or as componentwise addition). The

---

[1] The letter j is used both for $\sqrt{-1}$ and as an index throughout the book. This should not cause any confusion.

definition of multiplication is more elaborate. First find an irreducible polynomial $p(x)$ of degree m over F, that is, a polynomial with coefficients in
F that cannot be factored into a product of two polynomials of smaller degree with coefficients in F. (It is conventional to choose $p(x)$ to be a *monic
polynomial*, defined as a polynomial with leading coefficient equal to 1.)
Then multiplication is defined to be polynomial multiplication modulo the
polynomial $p(x)$.

$$c(x) = a(x)b(x) \quad (\text{mod } p(x)),$$

where the reduction modulo the polynomial $p(x)$ simply means replacing a
polynomial by its remainder when divided by $p(x)$. In this way, a polynomial product is collapsed back to a polynomial of degree at most $m-1$, so
the set F^m is closed under multiplication.

This construction always gives a field of the same characteristic as F
because the element 1 behaves the same under addition with itself whether
it is considered an element of F or of the extension field F^m. The structure of the field will depend on the choice of $p(x)$. For this reason, the
full name for the extension field is, by convention, $F/(p(x))$. Thus, by this
convention, the complex field would also be called $\mathbf{R}/(x^2 + 1)$. Sometimes,
we will be lax and simply call the field F^m, though this notation does not
fully specify the field because there are multiple possibilities for the irreducible polynomial $p(x)$. Technically, for finite fields, choosing a different
irreducible polynomial of degree m does not give a different field but only
a different way of describing the same field. There is only one field F^m if F
is a finite field, but it may be described in different ways, called *isomorphic*
descriptions. From a computational point of view, two descriptions of the
same field may appear to be different fields and so may be called different
fields.

Irreducible polynomials of every degree do not exist over every field.
For example, there are no irreducible polynomials of degree 4 over the
real field, so we cannot define a field structure on $\mathbf{R}^4$. Over fields of finite
characteristic, however, irreducible polynomials of every degree do exist.

2.3.1 FIELDS OF CHARACTERISTIC 2

Finite fields of the form $GF(2^m)$ are called fields of characteristic two or,
more simply, binary fields. The elements of the field can be thought of as
polynomials of degree $m-1$ with binary coefficients or can be thought of
simply as m-bit binary numbers. Addition is componentwise modulo-2 addition (bit-by-bit exclusive-or). Multiplication is polynomial multiplication
modulo $p(x)$, an irreducible polynomial over $GF(2)$ of degree m. A list of
irreducible polynomials over $GF(2)$ of various degrees is as follows:

$$\begin{aligned} p(x) \ &= \ x^2 + x + 1 \\ &\quad x^3 + x + 1 \end{aligned}$$

$$x^4 + x + 1$$
$$x^5 + x^2 + 1$$
$$x^6 + x + 1$$
$$x^7 + x^3 + 1$$
$$x^8 + x^4 + x^3 + x^2 + 1$$
$$x^9 + x^4 + 1$$
$$x^{10} + x^3 + 1$$
$$x^{11} + x^2 + 1$$
$$x^{12} + x^6 + x^4 + x + 1$$
$$x^{13} + x^4 + x^3 + x + 1$$
$$x^{14} + x^{10} + x^6 + x + 1$$
$$x^{15} + x + 1$$
$$\vdots$$

These irreducible polynomials over $GF(2)$ have been chosen to have an additional useful property. In the field $GF(2^m)$, when constructed with the above polynomial of degree m, the element x is a primitive element; every nonzero element of $GF(2^m)$ is a power of x (mod $p(x)$). Irreducible polynomials of this kind are called *primitive polynomials*.

For example, we can construct $GF(8)$ as an extension of $GF(2)$ by taking powers of the primitive element $\alpha = x$. Modulo $p(x) = x^3 + x + 1$, we have $x^3 = x + 1$, so the first seven powers of α are

$$\alpha = x$$
$$\alpha^2 = x^2$$
$$\alpha^3 = x + 1$$
$$\alpha^4 = x^2 + x$$
$$\alpha^5 = x^2 + x + 1$$
$$\alpha^6 = x^2 + 1$$
$$\alpha^7 = 1.$$

This list implicitly contains the multiplication table for $GF(8)$. The exponent of α can be regarded as a logarithm of the three-bit binary number comprising the three coefficients of the field element α^i.

The field $GF(256)$ can be constructed as an extension of the field $GF(2)$ by using the polynomial

$$p(x) = x^8 + x^4 + x^3 + x^2 + 1.$$

This polynomial has coefficients in $GF(2)$ and cannot be factored in $GF(2)$, which can be verified by trying all possible factorizations. Then $GF(256)$

consists of all eight-bit binary numbers thought of as polynomials of degree at most 7 with coefficients in $GF(2)$. Addition is componentwise addition in $GF(2)$; that is, bit-by-bit exclusive-or. Multiplication consists of multiplication as polynomials reduced modulo the polynomial $p(x)$.

2.3.2 FIELDS THAT MIMIC THE COMPLEX FIELD

The field $GF(p^2)$ has a multiplication that mimics complex multiplication whenever $p = -1 \pmod 4$.

We can always extend $GF(p)$, p a prime, to the field $GF(p^2)$. This extension field will look familiar if the polynomial

$$p(x) = x^2 + 1$$

cannot be factored in $GF(p)$, which is equivalent to saying that $\sqrt{-1}$ does not exist in $GF(p)$. This, we have seen, is true for all primes satisfying $p = -1 \pmod 4$. Hence, the field elements are defined as the set of first-degree polynomials

$$GF(p^2) = \{a_1 x + a_0\}$$

with a_1 and a_0 elements of $GF(p)$, and arithmetic in $GF(p^2)$ is defined as polynomial addition and polynomial multiplication modulo the polynomial $x^2 + 1$. This construction gives a field. The structure will appear even more familiar if we use j in place of x and write the field as

$$GF(p^2) = \{a + jb\}$$

with addition given by

$$
\begin{aligned}
(a + jb) + (c + jd) &= (a + c) + j(b + d), \\
(a + jb)(c + jd) &= (ac - bd) + j(ad + bc).
\end{aligned}
$$

This is the same structure as the complex field, but of course the additions and multiplications on the right are in $GF(p)$. They are integer additions and multiplications modulo p.

For example, $\sqrt{-1}$ does not exist in the field $GF(2^{17} - 1)$, so we can construct the extension field

$$GF((2^{17} - 1)^2) = \{a + jb\},$$

where a and b are elements of $GF(2^{17} - 1)$. Addition and multiplication in this field act much as they do in the complex field, but the "real part" and "imaginary part" are integers under modulo $2^{17} - 1$ arithmetic.

2.3.3 FIELDS THAT CANNOT MIMIC THE COMPLEX FIELD

The field $GF(p^2)$ does not have a multiplication that mimics complex multiplication whenever $p = 1 \pmod 4$.

Whenever $\sqrt{-1}$ does exist in a field, then the polynomial $x^2 + 1$ is not irreducible, so it cannot be used to extend the field. For example, $\sqrt{-1}$ does exist in $GF(2^{16} + 1)$ given by $\sqrt{-1} = 2^8$ because $(2^8)^2 + 1 = 0$ (modulo $2^{16} + 1$). Therefore, $\sqrt{-1}$ already exists. It cannot be appended artificially, so $GF(2^{16} + 1)$ cannot be extended to a field that behaves like the complex field. It may be instructive, however, to realize that $GF(2^{16} + 1)$ can be extended to $GF((2^{16}+1)^2)$ even though the polynomial $p(x) = x^2 + 1$ is not irreducible. This is because there is another polynomial $p(x) = x^2 + x + 1$ that is irreducible in $GF(2^{16} + 1)$. Using this polynomial, multiplication in $GF((2^{16} + 1)^2)$ is given by

$$(a + jb)(c + jd) = (ac - bd) + j(ad + bc - bd),$$

where we have used j instead of x to emphasize the contrast with multiplication in the complex field. The extension field $GF((2^{16} + 1)^2)$ does exist, but its multiplication does not mimic complex multiplication.

Computations that involve multiplication of complex integers cannot be readily embedded into $GF((2^{16} + 1)^2)$ for processing. There is a compensating advantage, however. Because $GF(2^{16} + 1)$ already contains $\sqrt{-1}$, some computational problems involving complex numbers can be executed more simply in $GF(2^{16} + 1)$ and in such a way that the answers as complex integers can be recovered from the answers in $GF(2^{16} + 1)$. Specifically, we shall see in Chapter 5 how complex convolutions can be computed by embedding them in $GF(2^{16} + 1)$ and using "real" convolutions there.

2.3.4   FIELDS OF RATIONAL FUNCTIONS

There are also fields with finite characteristic but an infinite number of elements. A simple example is a *field of rational functions*.

Polynomials can be defined over any field F by

$$a(x) = \sum_{i=0}^{A} a_i x^i,$$

where a_i is an element of F and A is the degree of $a(x)$. (The degree of a polynomial is finite.) The set

$$F(x) = \left\{ \frac{a(x)}{b(x)} \mid b(x) \neq 0, \quad \mathrm{GCD}[a(x), b(x)] = 1 \right\},$$

where $a(x)$ and $b(x)$ are polynomials over F, forms a field with the usual definitions of addition and multiplication of rationals. The field of rationals over F has an infinite number of elements but has the same characteristic as the base field F.

2.3.5   CYCLOTOMIC EXTENSIONS OF THE RATIONAL FIELD

Most digital signal processing takes place within a small set of numbers —
say, the set of 16-bit fixed-point numbers. On the other hand, most of the
theoretical developments are carried out in a very rich number system —
the field of complex numbers. Numbers such as $\sqrt{2}$ never exist in computations; only approximations exist, but $\sqrt{2}$ is a quite ordinary number within
the theory of real or complex numbers.

There are many other fields that lie strictly between the rational field $\mathbf{Q}$
and the complex field $\mathbf{C}$. These fields are extensions of the rational field.
They have attracted the engineer's attention only rarely in the past, but
eventually the engineer turns everything he encounters to some practical
purpose.

It is of course very beneficial that we have the elegance and power of the
complex number system available to us for many purposes. However, the
complex number system includes many numbers that the practical computations cannot represent exactly and we usually blame this on limited
precision of the practical computations rather than on too high an expec-
tation built into the theory.

Let $x^n - 1$ be factored into its (monic) irreducible factors over the rationals.

$$x^n - 1 = p_0(x)p_1(x)\ldots p_s(x).$$

The irreducible factors of $x^n - 1$ over $\mathbf{Q}$ are known as *cyclotomic polyno-*
mials. These polynomials always have integer coefficients, and when n is
small, they have coefficients only equal to -1, 0, or $+1$. The cyclotomic
polynomials have the property that $x^n - 1$ has only one factor that is not
also a factor of $x^{n'} - 1$ for some n' smaller than n. This leads to a nice
way to index the cyclotomic polynomials. The polynomial $\Phi_n(x)$ is that
factor of $x^n - 1$ that is not a factor of $x^{n'} - 1$ for any n' less than n. Some
cyclotomic polynomials are as follows:

$$
\begin{aligned}
\Phi_1(x) &= x - 1 \\
\Phi_2(x) &= x + 1 \\
\Phi_3(x) &= x^2 + x + 1 \\
\Phi_4(x) &= x^2 + 1 \\
\Phi_5(x) &= x^4 + x^3 + x^2 + x + 1 \\
\Phi_6(x) &= x^2 - x + 1 \\
\Phi_7(x) &= x^6 + x^5 + x^4 + x^3 + x^2 + x + 1 \\
\Phi_8(x) &= x^4 + 1.
\end{aligned}
$$

For example, when $n = 16$,

$$
\begin{aligned}
x^{16} - 1 &= (x^8 + 1)(x^4 + 1)(x^2 + 1)(x + 1)(x - 1) \\
&= \Phi_{16}(x)\Phi_8(x)\Phi_4(x)\Phi_2(x)\Phi_1(x),
\end{aligned}
$$

and when $n = 10$,

$$
\begin{aligned}
x^{10} - 1 &= (x^4 - x^3 + x^2 - x + 1)(x^4 + x^3 + x^2 + x + 1)(x + 1)(x - 1) \\
&= \Phi_{10}(x)\Phi_5(x)\Phi_2(x)\Phi_1(x).
\end{aligned}
$$

Notice that $\Phi_r(x)$ divides $x^n - 1$ if r divides n. This is true in general.

Let $\omega = e^{-j2\pi/n}$ be an nth root of unity. Because ω is a zero of $x^n - 1$, it is a zero of one of the cyclotomic polynomials. Because ω is not a zero of $x^{n'} - 1$ for $n' < n$, it must be a zero of $\Phi_n(x)$, a polynomial with leading coefficient equal to 1:

$$
\Phi_n(x) = x^m + p_{m-1}x^{m-1} + \ldots + p_1 x + p_0,
$$

where m is the degree of $\Phi_n(x)$.

Because all powers of ω are zeros of $x^n - 1$, all powers of ω that are not zeros of $x^{n'} - 1$ for $n' < n$ are zeros of $\Phi_n(x)$. Therefore,

$$
\Phi_n(x) = \prod_{\mathrm{GCD}(k,n)=1} (x - \omega^k)
$$

and

$$
\deg \Phi_n(x) = \phi(n).
$$

Because $\Phi_n(\omega) = 0$, the cyclotomic polynomial gives

$$
\omega^m = -p_{m-1}\omega^{m-1} - \ldots - p_1\omega - p_0.
$$

Hence ω^m can be expressed as a linear combination of lesser powers of ω. However, if i is less than m, then ω^i cannot be so expressed because if it could, ω would be a zero of another polynomial of degree smaller than the degree of $p(x)$.

The extension field $\mathbf{Q}(\omega)$ (or $\mathbf{Q}^m$) can be represented as the set of polynomials in ω with degree $m - 1$ or less and with rational coefficients. Each polynomial is represented by a list of m coefficients. Addition is polynomial addition, and multiplication is polynomial multiplication modulo $p(\omega)$. In a practical implementation, it takes m words of memory to store one element of $\mathbf{Q}(\omega)$ instead of taking two words of memory, which suffices for the usual complex numbers.

In order to emphasize that the number representation consists of the polynomials themselves and not the complex value the polynomial takes at ω, we use the variable x in place of ω. Then the numbers are

$$
a = a_{m-1}x^{m-1} + a_{m-2}x^{m-2} + \ldots + a_1 x + a_0,
$$

as represented by the list of coefficients. Of course, if we wanted to embed $\mathbf{Q}^m$ into $\mathbf{C}$ to recover the "true" complex value of a, we could just substitute ω for x and carry out the indicated calculations. However, our aim is to derive algorithms that use the polynomial representation for intermediate variables, and this form will actually lead to simpler algorithms in some instances.

For example, if n is a power of 2, then

$$x^n - 1 = (x^{n/2} + 1)(x^{n/4} + 1) \ldots (x + 1)(x - 1).$$

The cyclotomic polynomial $x^{n/2} + 1$ leads to an extension field whose elements are all the polynomials with rational coefficients of degree less than $n/2$; addition is polynomial addition; multiplication is polynomial multiplication modulo the polynomial $x^{n/2} + 1$.

More specifically, if n is 16, then $p(x) = x^8 + 1$ and the extension field $\mathbf{Q}^8$ consists of all rational-valued polynomials of degree seven or less. Multiplication is polynomial multiplication modulo the polynomial $x^8 + 1$. A sample multiplication in the field $\mathbf{Q}^8$ is

$$
\begin{aligned}
\left(x^7 - \frac{1}{2}x^2 + 1\right)(x^2 - 1) &= x^9 - x^7 - \frac{1}{2}x^4 + \frac{3}{2}x^2 - 1 \quad (\text{mod } x^8 + 1) \\
&= -x^7 - \frac{1}{2}x^4 + \frac{3}{2}x^2 - x - 1.
\end{aligned}
$$

A number in the field $\mathbf{Q}^8$ can be thought of as a kind of generalization of a complex number having eight parts; each part must have a rational value. In this field (using j instead of x as a place marker), the relation $j^8 = -1$ replaces the familiar $j^2 = -1$ of the complex field.

It may be desired in some applications to have the usual complex rationals included as a subfield of the extension field $\mathbf{Q}(\omega)$. If the extension field $\mathbf{Q}(\omega)$ contains j (that is, contains a zero of the polynomial $y^2 + 1$), then this extension field already contains all the complex rationals; it need not be enlarged. The field will contain j if the irreducible polynomial is $x^n + 1$ with n even. In that case $j = x^{n/2}$ because then $j^2 + 1 = 0$ modulo the polynomial $x^n + 1$. Otherwise, the element j must be appended if the field is to be used for computations with complex rationals.

Let ω be an nth root of unity, n not a power of 2, and let $\Phi_n(x)$ be the cyclotomic polynomial of degree m with ω as a zero. Then the extension field $\mathbf{Q}(j, \omega)$ is the set of polynomials of degree less than m with coefficients in $\mathbf{Q}(j)$. Addition is polynomial addition; multiplication is polynomial multiplication modulo the polynomial $p(x)$. Coefficients of the polynomials are added and multiplied as complex numbers. It requires $2m$ rational numbers to specify one element of $\mathbf{Q}(j, \omega)$.

2.4 Rings and Groups

A ring is an algebraic system that is weaker than a field in that division is not necessarily defined for all nonzero elements. A group is weaker still in that only a single operation is defined.

A formal definition of a ring is as follows. A *ring R* is a set that has two arithmetic operations defined on it: addition and multiplication, such that the following properties are satisfied.

1) (Addition Axiom) The set R is closed under addition, and addition is associative and commutative

$$a + (b + c) = (a + b) + c,$$
$$a + b = b + a.$$

There is an element called *zero* and denoted 0 such that $a + 0 = a$, and every element a has an element called the *negative* of a and denoted $(-a)$ such that $a + (-a) = 0$. Subtraction $a - b$ is defined as $a + (-b)$.

2) (Multiplication Axiom) The set R is closed under multiplication, and multiplication is associative

$$a(bc) = (ab)c.$$

3) (Joint Axiom) The distributive laws

$$(a + b)c = ac + bc,$$
$$c(a + b) = ca + cb$$

hold for all numbers a, b, and c in the set R.

A ring need not have an identity under multiplication but if it does that element is denoted 1, and called *one*, and the ring is called a *ring with identity*. The multiplication operation in a ring need not be commutative but if it is, the ring is called a *commutative* ring.

The most familiar example of a ring is the set of integers, denoted **Z**, and given by $\mathbf{Z} = \{0, \pm 1, \pm 2, \ldots\}$ with the usual definition of addition and multiplication. Notice that in **Z**, division is not defined in general but is sometimes meaningful (as in 6 divided by 3). This is also the case in other rings. Although division is not defined in general, for some a the equation $ab = 1$ may have a solution b which is called the *inverse* of a. Moreover, for some a and c, the equation $ab = c$ may have a solution for b even though a has no inverse.

The set of nonnegative integers smaller than q under modulo q arithmetic is called the ring of integers modulo q and is denoted by $\mathbf{Z}_q$. It is a field if and only if q is a prime. Thus

$$\mathbf{Z}_p = GF(p),$$
$$\mathbf{Z}_{p^m} \neq GF(p^m).$$

A group is the simplest (?) algebraic structure that we shall consider. A group has only a single operation defined on it.

A formal definition of a group is as follows. A *group* G is a set together with an operation, denoted $*$, on pairs of elements satisfying four properties:

1) (Closure)

    $c = a * b$ is in the group.

2) (Associativity)

    $a * (b * c) = (a * b) * c$.

3) (Identity) e exists in G such that for all a,

    $a * e = e * a = a$.

4) (Inverses) For every a, there exists b such that

    $a * b = b * a = e$.

A group G with a finite number of elements is called a *finite group*. In a finite group, ω has *order* n if n is the least positive integer such that $\omega^n = e$, where $\omega^n = \omega * \omega * \ldots * \omega$ with n copies of ω on the right. Every element of a finite group has a well-defined order. The order of any element ω of a finite group divides the number of elements in the group. To prove this, form the array

$$
\begin{array}{cccccc}
\omega & \omega^2 & \omega^3 & \cdots & \omega^{n-1} & \omega^n = 1 \\
g_2 * \omega & g_2 * \omega^2 & g_2 * \omega^3 & \cdots & g_2 * \omega^{n-1} & g_2 * \omega^n \\
g_3 * \omega & g_3 * \omega^2 & g_3 * \omega^3 & \cdots & g_3 * \omega^{n-1} & g_3 * \omega^n \\
\vdots & & & & & \\
g_m * \omega & g_m * \omega^2 & g_m * \omega^3 & \cdots & g_m * \omega^{n-1} & g_m * \omega^n
\end{array}
$$

as follows. Write the powers of ω in the first row. In forming the ith row choose any g_i that does not appear in any previous row and place $g_i * \omega^j$ in the ith row and jth column. Stop forming new rows when no previously unused element g_i exists. The construction must stop because the group is finite.

No group element can appear twice in the same row because if $g_i * \omega^j = g_i * \omega^\ell$ then $\omega^j = \omega^\ell$, which cannot be. No group element can appear in two different rows because if $g_i * \omega^j = g_k * \omega^\ell$ with $i < k$, then $g_k = g_i * \omega^{j-\ell}$, contrary to the condition that g_k not appear earlier in the array.

Hence every element of G appears exactly once in the rectangular array and the number of columns divides the number of elements of the array. Therefore the order of ω divides the number of elements in G.

An *abelian (or commutative) group* is a group that satisfies the commutative law

$$a * b = b * a$$

for all elements of the group. In an abelian group the group operation is often called "addition" and denoted by $+$. Then the inverse is called "zero" and denoted by 0.

A field contains a group structure in two ways. The field elements form an abelian group under addition and the nonzero field elements form an abelian group under multiplication. An element of a finite field has an order under addition and a nonzero element of a finite field has an order under multiplication. Unless the additive order is explicitly mentioned, the term "order" in a finite field refers to multiplicative order. The finite field $GF(q)$ has $q - 1$ elements in its multiplicative group. Hence the order of every nonzero element of $GF(q)$ divides $q - 1$.

2.5   Algebraic Integers

Let ω be a primitive nth root of unity in the complex field with $n \geq 3$. It is a zero of the cyclotomic polynomial

$$\Phi_n(x) = \prod_{\text{GCD}(k,n)=1} (x - \omega^k),$$

where the product is over all k less than n and relatively prime to n. The degree of the polynomial $\Phi_n(x)$ is equal to $\phi(n)$.

A cyclotomic extension of the rational field has been defined as the set

$$\mathbf{Q}(\omega) = \mathbf{Q}^m = \{a_{m-1}\omega^{m-1} + a_{m-2}\omega^{m-2} + \ldots + a_1\omega + a_0\},$$

where ω is an nth root of unity and the a_i are elements of $\mathbf{Q}$. We write the elements of $\mathbf{Q}^m$ in the form of polynomials in ω but, when we wish, we can assign ω its numerical value in $\mathbf{C}$ and so map the elements of $\mathbf{Q}^m$ into $\mathbf{C}$.

The field $\mathbf{Q}(\omega)$ contains the subset

$$\mathbf{Z}[\omega] = \{a_{m-1}\omega^{m-1} + a_{m-2}\omega^{m-2} + \ldots + a_1\omega + a_0 \mid a_i \in \mathbf{Z}\}.$$

Now the coefficients are integers. The set $\mathbf{Z}[\omega]$, known as a set of *algebraic integers*, is no longer a field, but it is a commutative ring with identity. The sum of two elements of $\mathbf{Z}[\omega]$ is an element of $\mathbf{Z}[\omega]$ and, because ω^m is an integer combination of smaller powers of ω, the product of two elements of $\mathbf{Z}[\omega]$ is an element of $\mathbf{Z}[\omega]$. In particular, if m is a power of 2, multiplication is simple because $\omega^m = -1$.

The ring $\mathbf{Z}[e^{j2\pi/4}]$, consisting of the set $\{a + jb\}$ where a and b are integers, is known as the ring of *Gaussian integers*. The Gaussian integers form a set of points on a square grid in the complex plane. An arbitrary point of the complex plane cannot be well approximated by a Gaussian integer. In contrast, the ring $\mathbf{Z}[e^{j2\pi/8}]$ is *dense* in the complex plane. This means that every complex number can be approximated arbitrarily precisely by an element of $\mathbf{Z}[e^{j2\pi/8}]$. Even the subset

$$\mathbf{Z}[e^{j2\pi/8}]_6 = \{a_7\omega^7 + a_6\omega^6 + \ldots + a_1\omega + a_0 \mid a_i \in \{-3, -2, -1, 0, 1, 2, 3\}\},$$

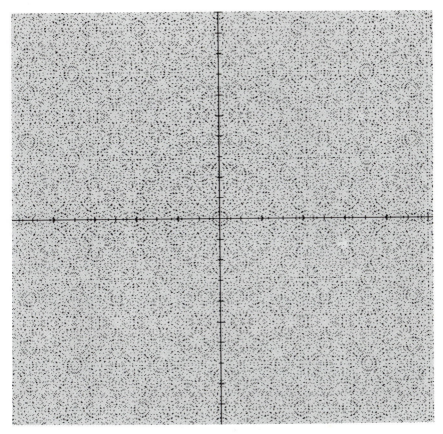

FIGURE 2.1. $\mathbf{Z}[e^{j2\pi/16}]_4$ near the origin.

in which the coefficients are restricted to be small (three-bit) integers, provides surprisingly good approximations near the origin of the complex plane. Figure 2.1 shows the set

$$\mathbf{Z}[e^{j2\pi/16}]_4 = \left\{a_{15}\omega^{15} + a_{14}\omega^{14} + \ldots + a_1\omega + a_0 \mid a_i \in \{-2, -1, 0, 1, 2\}\right\}$$

embedded in the complex plane; each point is represented by a dot. Figure 2.2 shows a similar set of points from $\mathbf{Z}[e^{j2\pi/16}]_6$ embedded in the complex plane.

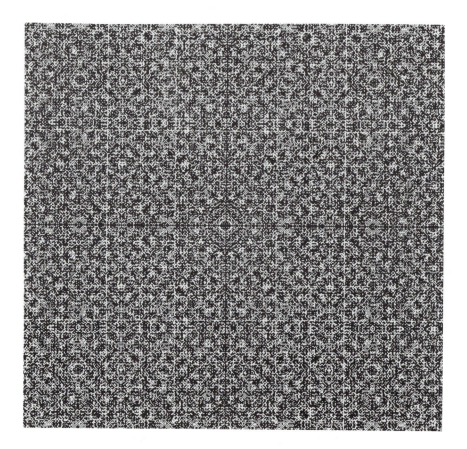

FIGURE 2.2. $\mathbf{Z}[e^{j2\pi/16}]_6$ near the origin.

3

Sequences and Spectra

Traditionally, digital signal processing deals with vectors whose components are elements of the complex number system or the real number system. The discrete Fourier transform is commonly defined in the complex field and it forms a fundamental tool in the subject of digital signal processing. Likewise, many of the tasks of error-control codes can be described in terms of the discrete Fourier transform in a Galois field. The rich set of properties of the Fourier transform — which the engineer has internalized to form part of his intuition — are valid in the complex field and in the Galois fields. These properties underlie much of the structure of digital signal processing and of error-control codes. Indeed, these properties hold in an arbitrary field and can be developed without mention of any particular field.

We shall define the discrete Fourier transform in an arbitrary field. Because we only consider the case where the index set is discrete (and finite), we shall refer to the discrete Fourier transform simply as the Fourier transform.

3.1   Weight and Complexity

Let F be a field and let $\mathbf{v} = (v_0, \ldots, v_{n-1})$ be a vector of blocklength n of numbers from the field F. We shall be interested in studying the structure and properties of such vectors. The two properties that we define in this section are the weight and the complexity. The *weight* of $\mathbf{v}$, denoted $w(\mathbf{v})$, is defined as the number of components in which $\mathbf{v}$ is nonzero. The weight of $\mathbf{v}$ is at most n.

The complexity is a little less straightforward to define. In general, to define the complexity of a sequence is not as straightforward as it may seem. Hence we deal only with those restricted concepts of complexity that are well-suited to our needs.

3.1.1 LINEAR COMPLEXITY

The *linear complexity* of the vector $\mathbf{v}$ is defined as the smallest value of t for which a recursion of the form

$$v_i = -\sum_{j=1}^{t} \Lambda_j v_{i-j}, \qquad i = t, \ldots, n-1$$

exists that will generate $\mathbf{v}$ from its first t components. The linear complexity of the all-zero sequence is defined to be zero. The linear complexity of a vector of length n is at most n.

The recursion

$$v_i = -\sum_{j=1}^{t} \Lambda_j v_{i-j}$$

can be thought of as a description of a linear feedback shift register (or autoregressive filter). As it is shifted, the linear feedback shift register initialized with $v_0, v_1, \ldots, v_{t-1}$ will generate all of $\mathbf{v}$. Accordingly, we may define the linear complexity of $\mathbf{v}$ as the length of the shortest linear feedback shift register that will generate all of $\mathbf{v}$ when initialized with the beginning of $\mathbf{v}$.

We can use the coefficients of the recursion or the tap weights of the linear feedback shift register to define a *feedback polynomial* $\Lambda(x)$ by

$$\begin{aligned}
\Lambda(x) &= 1 + \sum_{j=1}^{t} \Lambda_j x^j \\
&= \sum_{j=0}^{t} \Lambda_j x^j,
\end{aligned}$$

where for later convenience we have appended the extra coefficient $\Lambda_0 = 1$. The recursion of the linear feedback shift register is frequently denoted by the designation $(\Lambda(x), t)$. The reason for including both $\Lambda(x)$ and t in this designation is that Λ_t may equal zero, so the degree of $\Lambda(x)$ does not necessarily specify t.

In the rational field, the shift register $(\Lambda(x), t) = (x^2 + x + 1, 2)$ initialized with $(1, 1)$ generates the *Fibonacci sequence*, which is not periodic nor even eventually periodic. The linear complexity of the Fibonacci sequence — or any nontrivial segment of it — is 2 because it cannot be generated by any shorter linear feedback shift register. If, however, a new periodic sequence is formed by periodically repeating the first n symbols of the Fibonacci sequence, then that periodic sequence has linear complexity equal to at least $n - 1$. This is a consequence of Massey's theorem (to be proved in Section 6.2), which says that if $(v_0, v_1, \ldots, v_{n-1})$ has linear complexity t, then either $(v_0, v_1, \ldots, v_{n-1}, v_n)$ is generated by that same linear feedback shift register or $(v_0, v_1, \ldots, v_{n-1}, v_n)$ has linear complexity at least as large as $\max[t, n+1-t]$.

□ **Theorem 3.1.1** *The acyclic recursion*

$$v_i = -\sum_{j=1}^{t} \Lambda_j v_{i-j}, \qquad i = t, \ldots, n-1$$

can be expressed as the polynomial product

$$\Lambda(x)v(x) = p(x) + x^n g(x),$$

where $\Lambda_0 = 1$, $\deg p(x) < t$ and $g(x)$ is arbitrary.

Proof The acyclic recursion can be rewritten as the acyclic convolution

$$\sum_{j=0}^{t} \Lambda_j v_{i-j} = 0, \qquad i = t, \ldots, n-1,$$

where again $\Lambda_0 = 1$. The left side can be rewritten as the coefficients of a polynomial product $\Lambda(x)v(x)$. The coefficients of this polynomial product must be zero for $i = t, \ldots, n-1$.   □

A statement that is equivalent to the theorem is

$$\Lambda(x)v(x) = p(x) \pmod{x^n}.$$

In this form, the linear complexity of $v(x)$ is the smallest t for which this polynomial equation has a solution for $\Lambda(x)$ and $p(x)$ with $\deg \Lambda(x) \leq t$ and with $\deg p(x) < t$.

3.1.2 Cyclic Complexity

The *cyclic complexity* of the vector $\mathbf{v}$ having blocklength n is defined as the smallest value of t for which a cyclic recursion of the form

$$v_i = -\sum_{j=1}^{t} \Lambda_j v_{((i-j))}, \qquad i = t, \ldots, n+t-1$$

exists that will cyclically generate $\mathbf{v}$ from its first t components. The cyclic complexity of the all-zero sequence is defined to be zero.

The autoregressive filter of length n with $\Lambda_n = -1$, and otherwise $\Lambda_j = 0$, will cyclically generate any sequence of blocklength n because the shift register is initialized with $(v_0, v_1, \ldots, v_{n-1})$ and the shift register output is

$$v_i = v_{i-n}.$$

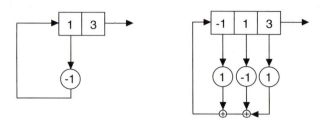

FIGURE 3.1. Two autoregressive filters to generate $(3, 1, -1, 1)$.

Therefore it is clear that the cyclic complexity of a sequence of length n is at most n.

The cyclic recursion

$$v_i = -\sum_{j=1}^{t} \Lambda_j v_{((i-j))}$$

holds for all $i \pmod{n}$ if it holds for $i = t, \ldots, n+t-1$. It can be rewritten as the cyclic convolution

$$\sum_{j=0}^{t} \Lambda_j v_{((i-j))} = 0, \qquad i = 0, \ldots, n-1,$$

where $\Lambda_0 = 1$. This cyclic convolution can be expressed as a polynomial product

$$\Lambda(x)v(x) = 0 \qquad (\bmod \ x^n - 1)$$

with

$$v(x) = \sum_{i=0}^{n-1} v_i x^i.$$

3.1.3   LINEAR AND CYCLIC COMPLEXITY

To end the section we consider the question of when the linear complexity and the cyclic complexity are equal (and so correspond to the same autoregressive filter). Figure 3.1 shows two autoregressive filters that generate the sequence $\mathbf{v} = (3, 1, -1, 1)$; one achieves the linear complexity and one achieves the cyclic complexity. These form an example which shows that the following theorem is not true if the term $\frac{n}{2}$ is made larger.

☐ **Theorem 3.1.2** *Let the characteristic of the field F be relatively prime to n (or infinite). The cyclic complexity and the linear complexity of a sequence in F of blocklength n are equal if the cyclic complexity is not larger than $\frac{n}{2}$.*

Proof Let $\overline{\Lambda}(x)$ achieve the linear complexity and let $\Lambda(x)$ achieve the cyclic complexity. Then $\deg \overline{\Lambda}(x) \leq \deg \Lambda(x)$, and we can write

$$\Lambda(x)v(x) = (x^n - 1)p(x),$$
$$\overline{\Lambda}(x)v(x) = (x^n - 1)\overline{p}(x) + \overline{r}(x),$$

where

$$\deg \overline{r}(x) < \deg \overline{\Lambda}(x) \leq \frac{n}{2}.$$

In a sufficiently large extension field $x^n - 1$ has n distinct zeros $\beta_1, \ldots, \beta_n$. Because $\deg \Lambda(x) \leq \frac{n}{2}$, $\Lambda(x)$ has at most $\frac{n}{2}$ zeros. This implies that $v(x)$ has at least $\frac{n}{2}$ zeros because

$$\Lambda(\beta_i)v(\beta_i) = 0$$

for all i.

Therefore $\overline{\Lambda}(x)v(x)$ has at least $\frac{n}{2}$ zeros among the zeros of $x^n - 1$. Furthermore, since

$$\overline{\Lambda}(x)v(x) = (x^n - 1)\overline{p}(x) + \overline{r}(x);$$

this implies that $\overline{r}(x)$ has at least $\frac{n}{2}$ zeros. But $\deg \overline{r}(x) < \frac{n}{2}$. Hence $\overline{r}(x)$ is the all-zero vector and the theorem is proved.   □

3.2   The Fourier Transform

We shall develop the Fourier transform and its properties in a way that is not dependent on the properties of a particular field F. This section and Section 3.4 should be read with the understanding that $\mathbf{v}$ is a vector over an arbitrary field F. Of course, the complex field can be used as a familiar example to help understand the many properties in Section 3.4 but the treatment is general and applies to any field.

Let $\mathbf{v} = (v_0, \ldots, v_{n-1})$ be a vector of blocklength n over the field F. The (discrete) Fourier transform of $\mathbf{v}$ is another vector $\mathbf{V}$ of blocklength n over the field F given by

$$V_k = \sum_{i=0}^{n-1} \omega^{ik} v_i, \qquad k = 0, \ldots, n - 1,$$

where ω is a number of order n in the field F. Thus $\omega^n = 1$ and no smaller power of ω equals 1. The Fourier transform looks quite familiar, but the operations of addition and multiplication that are indicated are operations in the field F. (The operations in the exponent of ω, however, are conventional integer arithmetic because ω^r means ω multiplied by itself r times in the field F.) We shall see that, in any field, all of the familiar properties of the Fourier transform are valid because as we look into the proof of these properties we find that only the abstract structure of a field is used; properties of addition and multiplication specific to a particular field are not used.

The only caution that must be made is that to have a Fourier transform of blocklength n in a field F, one must have an element ω of order n in F. In the complex field, for every n there is an element of order n given by

$$\omega = e^{-j2\pi/n},$$

where $j = \sqrt{-1}$. In other fields there need not exist an element of order n for every n. In particular, in the rational field or the real field we can only choose $\omega = -1$ and so can only have a Fourier transform of blocklength 2 in those fields. That Fourier transform is too small to be of much use. This is one reason why computational problems that start out in the real or rational field often are embedded into the complex field for processing.

The inverse Fourier transform is given next; it appears much the same as the Fourier transform except that ω^{-1} replaces ω, and that n^{-1} appears multiplying the sum. The term n^{-1} requires interpretation in an arbitrary field F. The term is the inverse of the integer of the field

$$n = 1 + \ldots + 1 \qquad (n \text{ terms}),$$

which is defined because 1 exists in F. It is worth noting that if $F = GF(2^m)$, the blocklength n must be odd; hence the sum of n 1's (modulo 2) equals 1. Consequently it is conventional to omit the $\frac{1}{n}$ in the inverse Fourier transform over $GF(2^m)$.

☐ **Theorem 3.2.1** *If $\mathbf{V}$ is the Fourier transform of $\mathbf{v}$, then $\mathbf{v}$ can be recovered from $\mathbf{V}$ by the inverse Fourier transform, which is given by*

$$v_i = \frac{1}{n} \sum_{k=0}^{n-1} \omega^{-ik} V_k.$$

Proof

$$\sum_{k=0}^{n-1} \omega^{-ik} V_k = \sum_{k=0}^{n-1} \omega^{-ik} \sum_{\ell=0}^{n-1} \omega^{\ell k} v_\ell$$

$$= \sum_{\ell=0}^{n-1} v_\ell \left[\sum_{k=0}^{n-1} \omega^{k(\ell-i)} \right].$$

But the sum on k is clearly equal to n if $\ell = i$, while if ℓ is not equal to i, then the sum on k becomes

$$\sum_{k=0}^{n-1} (\omega^{(\ell-i)})^k = \frac{1 - \omega^{(\ell-i)n}}{1 - \omega^{(\ell-i)}}.$$

The right side equals zero because $w^n = 1$. Hence

$$\sum_{k=0}^{n-1} w^{-ik} V_k = \sum_{\ell=0}^{n-1} v_\ell(n\delta_{i\ell}) = n v_i,$$

where $\delta_{i\ell} = 1$ if $i = \ell$ and otherwise $\delta_{i\ell} = 0$.   □

Sometimes we will write the Fourier transform as the matrix-vector product

$$\mathbf{V} = \mathbf{T}\mathbf{v},$$

where $\mathbf{T}$ is the matrix with elements w^{ik}. When written out explicitly, this equation becomes

$$
\begin{bmatrix} V_0 \\ V_1 \\ V_2 \\ \vdots \\ V_{n-1} \end{bmatrix}
=
\begin{bmatrix}
1 & 1 & 1 & \cdots & 1 \\
1 & w & w^2 & \cdots & w^{n-1} \\
1 & w^2 & w^4 & \cdots & w^{2(n-1)} \\
\vdots & \vdots & & & \\
1 & w^{n-1} & w^{2(n-1)} & \cdots & w
\end{bmatrix}
\begin{bmatrix} v_0 \\ v_1 \\ v_2 \\ \vdots \\ v_{n-1} \end{bmatrix}.
$$

3.3   Examples of Fourier Transforms

In this section we shall give numerous examples of Fourier transforms in a variety of fields. The examples are individually interesting for their own sake and also, by their number and variety, will emphasize that there is a considerable body of mathematical machinery here waiting to be applied.

1. **Q** or **R**. In the rational field or the real field an element w of order n exists only if n equals 1 or 2. This means that only trivial Fourier transforms exist in these fields. This is why one uses the extension field **C** to obtain Fourier transforms of real-valued vectors.

2. **C**. In the complex field an w of order n exists for every value of n. The conventional choice of w is $w = e^{-j2\pi/n}$, but other choices work. For example, $w = (e^{-j2\pi/n})^3$ works if n is not a multiple of 3.

3. $GF(5)$. In $GF(5)$ the element 2 has order 4 (because $2^4 = 1$ modulo 5). Therefore we have the four-point Fourier transform in $GF(5)$:

$$V_k = \sum_{i=0}^{3} 2^{ik} v_i, \qquad k = 0, \ldots, 3.$$

The components of $\mathbf{v}$ and $\mathbf{V}$ are elements of $GF(5)$ and all arithmetic is the arithmetic of $GF(5)$. If

$$\mathbf{v} = \begin{bmatrix} 1 \\ 2 \\ 3 \\ 4 \end{bmatrix}$$

then the Fourier transform of $\mathbf{v}$ is

$$\begin{bmatrix} V_0 \\ V_1 \\ V_2 \\ V_3 \end{bmatrix} = \begin{bmatrix} 1 & 1 & 1 & 1 \\ 1 & 2 & 4 & 3 \\ 1 & 4 & 1 & 4 \\ 1 & 3 & 4 & 2 \end{bmatrix} \begin{bmatrix} 1 \\ 2 \\ 3 \\ 4 \end{bmatrix} = \begin{bmatrix} 0 \\ 4 \\ 3 \\ 2 \end{bmatrix}.$$

4. $GF(31)$. In $GF(31)$, the element 2 has order 5. Therefore we have the five-point Fourier transform in $GF(31)$.

$$V_k = \sum_{i=0}^{4} 2^{ik} v_i, \qquad k = 0, \ldots, 4.$$

For example,

$$\begin{bmatrix} V_0 \\ V_1 \\ V_2 \\ V_3 \\ V_4 \end{bmatrix} = \begin{bmatrix} 1 & 1 & 1 & 1 & 1 \\ 1 & 2 & 4 & 8 & 16 \\ 1 & 4 & 16 & 2 & 8 \\ 1 & 8 & 2 & 16 & 4 \\ 1 & 16 & 8 & 4 & 2 \end{bmatrix} \begin{bmatrix} 2 \\ 6 \\ 4 \\ 4 \\ 0 \end{bmatrix} = \begin{bmatrix} 16 \\ 0 \\ 5 \\ 29 \\ 22 \end{bmatrix}.$$

5. $GF(2^m + 1)$. An element ω of order n exists if and only if n divides $2^m + 1 - 1$. Hence a Fourier transform of blocklength n exists if and only if $n = 2^r$ for some $r \leq m$. In such fields the Fourier transforms are also called Fermat number transforms.

6. $GF(2^{13} - 1)$. Because $2^{13} - 2 = 2 \cdot 5 \cdot 7 \cdot 9 \cdot 13$, only those values of n obtained by multiplying a selection of these factors can be used as a Fourier transform blocklength. It is easy to see that -2 has order 26 because $2^{13} = 1 \pmod{2^{13} - 1}$. Therefore $\omega = -2$ gives a Fourier transform with blocklength 26. Fourier transforms in $GF(2^m - 1)$ are also called Mersenne number transforms.

7. $GF(40961)$. Because $40960 = 2^{13} \cdot 5$, $GF(40961)$ has a Fourier transform of blocklength 2^{13}. A Fourier transform whose blocklength is a power of 2 is sometimes called a *radix-2* Fourier transform, especially if the components are thought of as indexed by binary numbers.

8. $GF((2^m - 1)^2)$. Because $(2^m - 1)^2 - 1 = 2^{m+1}(2^{m-1} - 1)$ this field has a Fourier transform of blocklength $n = 2^{m+1}$. Thus, although $GF(2^{17} - 1)$ has no Fourier transforms with power of 2 blocklength, the field $GF((2^{17}-1)^2)$ has such Fourier transforms up to blocklength $n = 2^{18}$. A vector over $GF(2^{17} - 1)$ of blocklength 2^{18} can be embedded in $GF((2^{17} - 1)^2)$ to get a Fourier transform of that blocklength. The element $\omega = 1021 + j4680$ has order 4096 in $GF((2^{17} - 1)^2)$. Therefore

$$V_k = \sum_{i=0}^{4095} (1021 + j4680)^{ik} v_i$$

is a Fourier transform. All arithmetic is modulo $2^{17} - 1$; that is, the overflow bit 2^{17} becomes a 1.

9. $GF((2^m + 1)^2)$. This field allows a Fourier transform of any block-length that divides $(2^m + 1)^2 - 1 = 2^{m+1}(2^{m-1} + 1)$. We can represent the elements of $GF((2^{16} + 1)^2)$ in the form $a + jb$, where a and b are elements of $GF(2^{16} + 1)$, and $j^2 = -j - 1$. Because j does not act the same as $\sqrt{-1}$ does in the complex field, we cannot embed complex integers into $GF((2^{16} + 1)^2)$ but we can embed real integers into the real part of $GF((2^{16} + 1)^2)$. Therefore Fourier transforms in $GF((2^{16} + 1)^2)$ can be used to convolve sequences of real integers in the same way that Fourier transforms in the complex field can be used to convolve sequences of real integers. Even though $\mathbf{v}$ may begin as a vector in $GF(2^{16} + 1)$, the transform $\mathbf{V}$ will be a vector in $GF((2^{16} + 1)^2)$. The spectral products have the complexity of complex multiplications though with a somewhat different structure.

10. $GF(256)$. Because $255 = 3 \cdot 5 \cdot 17$, this field has Fourier transforms of blocklengths 255, 85, 51, 17, 15, 5, and 3 and only Fourier transforms of these blocklengths. The Fourier transform of blocklength 255 takes a vector of 255 eight-bit bytes and maps it into a vector of 255 eight-bit bytes.

11. $\mathbf{Q}^{32}$. The extension of the rationals generated by the cyclotomic polynomial $p(x) = x^{32} + 1$ has elements consisting of polynomials of degree 31 or less. Because $x^{32} = -1 \pmod{x^{32} + 1}$, the element $\omega = x$ has order 64. Therefore, in this field, there is a Fourier transform of blocklength 64. We can write it explicitly as

$$V_k(x) = \sum_{i=0}^{63} x^{ik} v_i(x) \qquad (\bmod \ x^{32} + 1), \quad k = 0, \dots, 63.$$

The transform maps a vector of polynomials over $\mathbf{Q}$ into a vector of polynomials over $\mathbf{Q}$. Multiplication in $\mathbf{Q}^{32}$ by x or a power of x does not require any nontrivial multiplication in the base field $\mathbf{Q}$.

3.4   Properties of the Fourier Transform

The strong properties of the Fourier transform make it the important tool that it is. A list of the major properties is given in Figure 3.2. All of the properties hold in any field F; all of the properties except the last few are well known.

The linearity property is a trivial consequence of the definition, and the inverse transform was discussed in Section 3.1. The translation property is the dual of the modulation property.

1. Linearity

$$\lambda \mathbf{v} + \mu \mathbf{v}' \longleftrightarrow \lambda \mathbf{V} + \mu \mathbf{V}'$$

2. Inverse

$$v_i = \frac{1}{n} \sum_{k=0}^{n-1} \omega^{-ik} V_k \qquad i = 0, \ldots, n-1$$

$$\text{where } n = 1 + 1 + \ldots + 1 \qquad (n \text{ terms})$$

3. Modulation

$$(v_i \omega^{i\ell}) \longleftrightarrow (V_{((k+\ell))})$$

4. Translation

$$(v_{((i-\ell))}) \longleftrightarrow (V_k \omega^{\ell k})$$

5. Convolution Theorem

$$e_i = \sum_{\ell=0}^{n-1} f_{((i-\ell))} g_\ell \qquad i = 0, \ldots, n-1$$

if and only if

$$E_k = F_k G_k \qquad k = 0, \ldots, n-1$$

6. $v(x) = \sum_{i=0}^{n-1} v_i x^i$ has a zero at ω^k if and only if $V_k = 0$

7. The weight of a vector $\mathbf{v}$ is equal to the cyclic complexity of its Fourier transform $\mathbf{V}$.

FIGURE 3.2. Properties of the Fourier transform.

The modulation property is verified as follows: The transform of $v_i' = v_i \omega^{i\ell}$ is

$$
\begin{aligned}
V_k' &= \sum_{i=0}^{n-1} \omega^{ik} v_i' \\
&= \sum_{i=0}^{n-1} \omega^{i(k+\ell)} v_i \\
&= V_{k+\ell}.
\end{aligned}
$$

The *convolution theorem* is an important and familiar link between the Fourier transform and the cyclic convolution. We shall prove the convolution theorem here to make it quite clear that we are not restricted to any particular field when using it. The cyclic convolution of two vectors **f** and **g**, denoted **f** * **g**, is defined as

$$
e_i = \sum_{\ell=0}^{n-1} f_{((i-\ell))} g_\ell, \qquad i = 0, \ldots, n-1.
$$

☐ **Theorem 3.4.1** *The vector* **e** *is given by the cyclic convolution of the vectors* **f** *and* **g** *if and only if the components of the Fourier transforms satisfy*

$$
E_k = F_k G_k, \qquad k = 0, \ldots, n-1.
$$

Proof This holds because

$$
\begin{aligned}
e_i &= \sum_{\ell=0}^{n-1} f_{((i-\ell))} \left[\frac{1}{n} \sum_{k=0}^{n-1} \omega^{-k\ell} G_k \right] \\
&= \frac{1}{n} \sum_{k=0}^{n-1} \omega^{-ik} G_k \left[\sum_{\ell=0}^{n-1} \omega^{(i-\ell)k} f_{((i-\ell))} \right] = \frac{1}{n} \sum_{k=0}^{n-1} \omega^{-ik} G_k F_k.
\end{aligned}
$$

Because **e** is the inverse Fourier transform of **E**, we conclude that $E_k = G_k F_k$. ☐

3.4.1 Cyclic Decimation

If b and n are relatively prime, then $i \to bi \pmod{n}$ defines a permutation of the components of a vector **v**. The permutation $v_i' = v_{((bi))}$ is called *cyclic decimation* because every bth component, cyclically, is chosen.

☐ **Theorem 3.4.2** *Let* $\mathrm{GCD}(b, n) = 1$ *and* $Bb = 1 \pmod{n}$. *The cyclic decimation*

$$
v_i' = v_{((bi))}
$$

has transform with components

$$V_k' = V_{((Bk))},$$

where all indices are interpreted modulo n.

Proof The corollary to the Euclidean algorithm implies that

$$Bb + Nn = 1$$

for some integers B and N. Hence the required B exists. Therefore,

$$V_k' = \sum_{i=0}^{n-1} \omega^{ik} v_i'$$

$$= \sum_{i=0}^{n-1} \omega^{(Bb+Nn)ik} v_{((bi))}$$

$$= \sum_{i=0}^{n-1} \omega^{biBk} v_{((bi))}.$$

Because $i' = ((bi))$ is just a permutation, it does not affect the sum. Thus

$$V_k' = \sum_{i'=0}^{n-1} \omega^{i'Bk} v_{i'}$$

$$= V_{((Bk))},$$

as was to be proved. □

If $\mathrm{GCD}(b,n) \neq 1$, then the cyclic decimation $v_i' = v_{((bi))}$ has period smaller than n — the period is $n' = n/\mathrm{GCD}(b,n)$. The following theorem relates the n'-point Fourier transform of the decimated vector $\mathbf{v}'$ to the n-point Fourier transform of the original vector $\mathbf{v}$ by first folding the spectrum $\mathbf{V}$ then cyclically decimating the folded spectrum.

□ **Theorem 3.4.3** *Let $\mathrm{GCD}(b,n) = n''$, $n' = n/n''$, and $b' = b/n''$. The cyclic decimation*

$$v_i' = v_{((bi))}, \qquad i = 0, \ldots, n' - 1$$

has Fourier transform

$$V_{k'}' = \overline{V}_{B'k' \pmod{n'}}, \qquad k' = 0, \ldots, n' - 1,$$

where

$$\overline{V}_{k'} = \frac{1}{n''} \sum_{k''=0}^{n''-1} V_{k'+n'k''}, \qquad k' = 0, \ldots, n' - 1$$

and B' satisfies

$$B'b' = 1 \quad (\mathrm{mod}\ n').$$

Proof By the definition of $\text{GCD}(b, n)$, b' and n' can have no common factor. Hence $\text{GCD}(b', n') = 1$, so the designated B' exists.

The first step is to write $\mathbf{v}'$ in terms of $\mathbf{v}$, which in turn can be written as an inverse Fourier transform of $\mathbf{V}$. Thus

$$v'_i = v_{((bi))} = \frac{1}{n} \sum_{k=0}^{n-1} \omega^{-bik} V_k.$$

Write k in terms of a vernier index k' and a coarse index k'' as

$$k = k' + n'k'', \qquad k' = 0, \ldots, n' - 1, \quad k'' = 0, \ldots, n'' - 1.$$

Then

$$v'_i = \frac{1}{n} \sum_{k'=0}^{n'-1} \sum_{k''=0}^{n''-1} \omega^{-bik'} \omega^{-n'bik''} V_{k'k''},$$

where $V_{k'k''} = V_{k'+n'k''}$. Because $b = b'n''$ and $\omega^{n'n''} = 1$, the second term in ω can be replaced by 1. Then

$$v'_i = \frac{1}{n'} \sum_{k'=0}^{n'-1} \omega^{-n''b'ik'} \left[\frac{1}{n''} \sum_{k''=0}^{n''-1} V_{k'k''} \right].$$

Next replace k' by $B'k' \pmod{n'}$. This is a permutation of the indices $(0, 1, \ldots, n' - 1)$ so it does not affect the sum. Then

$$v'_i = \frac{1}{n'} \sum_{k'=0}^{n'-1} \omega^{-n''b'B'ik'} \left[\frac{1}{n''} \sum_{k''=0}^{n''-1} V_{B'k',k''} \right].$$

Because $b'B' = 1 \pmod{n'}$, this becomes

$$v'_i = \frac{1}{n'} \sum_{k'=0}^{n'-1} \gamma^{-ik'} \left[\frac{1}{n''} \sum_{k''=0}^{n''-1} V_{B'k',k''} \right],$$

where $\gamma = \omega^{n''}$. The proof is complete. $\square$

For example, suppose $n = 8$ and $b = 6$, and the decimated vector is

$$\mathbf{v}' = (v_0, v_6, v_4, v_2).$$

Then $n' = 4$, $n'' = 2$, $b' = 3$, and $B' = 3$. The theorem then gives the Fourier transform of the decimated vector as

$$\mathbf{V}' = \tfrac{1}{2}(V_0 + V_4, V_3 + V_7, V_2 + V_6, V_1 + V_5).$$

3.4.2   THE CYCLIC COMPLEXITY PROPERTY

A vector $\boldsymbol{\Lambda}$ of length t can be used to define the taps of a linear feedback shift register. The linear feedback shift register will cyclically generate the vector $\mathbf{V}$ if

$$V_k = - \sum_{j=0}^{t} \Lambda_j V_{((k-j))}.$$

Consider a vector $\mathbf{v}$ of weight 1 and its Fourier transform

$$V_k = v_m \omega^{km},$$

where v_m is the magnitude of the single nonzero component of $\mathbf{v}$. This spectrum can be generated by the recursion

$$V_k = \omega^m V_{k-1}$$

with the initial value $V_0 = v_m$. Thus the spectrum of a vector of weight 1 can be generated by a linear feedback shift register of length 1. This is a special case of the following theorem.

□ **Theorem 3.4.4** *The weight of a vector $\mathbf{v}$ is equal to the cyclic complexity of its Fourier transform $\mathbf{V}$.*

   Proof Recall that $(\Lambda(x), t)$ will cyclically generate $\mathbf{V}$ if and only if

$$\Lambda(x) V(x) = 0 \quad (\mathrm{mod}\ x^n - 1).$$

This is the cyclic convolution $\boldsymbol{\Lambda} * \mathbf{V} = \mathbf{0}$. Consequently, $\Lambda(x)$ will cyclically generate $V(x)$ if and only if

$$\lambda_i v_i = 0, \qquad i = 0, \ldots, n-1,$$

where $\boldsymbol{\lambda}$ and $\mathbf{v}$ are the inverse Fourier transforms of $\boldsymbol{\Lambda}$ and $\mathbf{V}$, respectively.
   The degree of $\Lambda(x)$ is not less than the number of its zeros and λ_i must be zero whenever v_i is nonzero. Therefore the feedback polynomial of smallest degree is the polynomial $\Lambda(x)$ that has a zero at ω^{-i} if and only if v_i is nonzero. The degree of this polynomial is equal to the weight of $\mathbf{v}$.   □

□ **Corollary 3.4.5** *The cyclic complexity of a vector $\mathbf{v}$ is equal to the weight of its Fourier transform.*   □

   From the proof of the theorem, it is clear that if $(\Lambda(x), t)$ generates the spectrum $\mathbf{V}$ then $\Lambda(x)$ has a simple zero at ω^{-i} if and only if $v_i \neq 0$. For this reason the minimal-length feedback polynomial $\Lambda(x)$ is also called the *locator polynomial* for $\mathbf{V}$; the zeros of $\Lambda(x)$ point to the nonzero components of the inverse Fourier transform of $\mathbf{V}$.

Let $i_1, i_2, \ldots, i_t$ index the t nonzero components of $\mathbf{v} = (v_0, v_1, \ldots, v_{n-1})$. Let

$$\Lambda(x) = \prod_{\ell=1}^{d}(1 - x\omega^{i_\ell})$$

$$= \sum_{k=0}^{t}\Lambda_k x^k.$$

The vector $\boldsymbol{\lambda}$ then satisfies

$$\lambda_i = \frac{1}{n}\sum_{k=0}^{n-1}\omega^{-ik}\Lambda_k$$

$$= \Lambda(\omega^{-i})$$

and $\lambda_i = 0$ if and only if $i \in \{i_1, \ldots, i_t\}$. That is, $\lambda_i = 0$ if and only if $v_i \neq 0$. Consequently, $\lambda_i v_i = 0$ for all i so the convolution theorem implies

$$\Lambda(x)V(x) = 0 \qquad (\mathrm{mod}\ x^n - 1)$$

and no polynomial of smaller degree can replace $\Lambda(x)$ in this equality.

3.4.3 Conjugacy Constraints

To obtain a Fourier transform of a vector of blocklength n over the field F it may be necessary to enter an extension field F^m. The Fourier transforms of real-valued vectors, for example, are complex valued. The vectors in F^m that are Fourier transforms of vectors in F satisfy a property known as a *conjugacy constraint*. The particular form of the conjugacy constraint depends on the field F.

The complex spectrum $\mathbf{V}$ is the Fourier transform of a real-valued vector if and only if $V_k^* = V_{n-k}$. This follows by writing

$$V_{n-k}^* = \left(\sum_{i=0}^{n-1}\omega^{i(n-k)}v_i\right)^*$$

$$= \sum_{i=0}^{n-1}\omega^{ik}v_i^*.$$

The asserted conjugacy constraint holds if and only if $v_i = v_i^*$ for all i.

The conjugacy constraint in fields of characteristic 2 is that the $GF(2^m)$-ary spectrum is the Fourier transform of a vector over $GF(2)$ if and only if

$$V_k^2 = V_{((2k))}.$$

This follows by writing

$$V_k^2 = \left(\sum_{i=0}^{n-1} \omega^{ik} v_i\right)^2$$

$$= \sum_{i=0}^{n-1} \omega^{2ik} v_i^2$$

because all other terms of the expansion contain a 2 and so are zero in fields of characteristic 2. If v_i is an element of $GF(2)$, then $v_i^2 = v_i$ and $V_k^2 = V_{((2k))}$. Conversely, if $V_k^2 = V_{((2k))}$ the uniqueness of the inverse Fourier transform implies that $v_i^2 = v_i$, so v_i is an element of $GF(2)$.

More generally, a vector $\mathbf{v}$ lies in the Galois field $GF(q)$ if and only if the transform in $GF(q^m)$ satisfies the conjugacy constraint

$$V_k^q = V_{((qk))}.$$

The proof is based on the relationship $(a + b)^p = a^p + b^p$ in fields of characteristic p, which holds because every other term in the expansion is a multiple of p and so equals zero modulo p. By repeated application of this rule, we conclude that

$$(a + b)^{p^m} = a^{p^m} + b^{p^m}$$

in fields of characteristic p. This allows us to write

$$V_k^q = \left(\sum_{i=0}^{n-1} \omega^{ik} v_i\right)^q$$

$$= \sum_{i=0}^{n-1} \omega^{qik} v_i^q.$$

Noting that $\mathbf{v}$ is a vector over $GF(q)$ if and only if $v_i^q = v_i$ leads to the conjugacy constraint.

For p satisfying $p = -1 \pmod 4$ the conjugacy constraint takes a more familiar form in the field $GF(p^2)$ constructed with the prime polynomial $j^2 + 1 = 0$. Then, by the conjugacy constraint, if $\mathbf{v}$ is in $GF(p)$,

$$V_k^p = V_{((pk))}.$$

But the left side can be expanded into a real part and an imaginary part as follows:

$$V_k^p = (V_{Rk} + jV_{Ik})^p$$

$$= V_{Rk}^p + j^p V_{Ik}^p$$

$$= V_{Rk} - jV_{Ik}$$

because V_{Rk} and V_{Ik} are each in $GF(p)$ (where all elements satisfy $\beta^p = \beta$) and $j^p = -j$ if $p = -1$ (mod 4). Therefore in $GF(p^2)$, with $p = -1$ (mod 4) the conjugacy constraint is

$$V_k^* = V_{((pk))},$$

which resembles the complex case.

The final case that we consider is the case of a cyclotomic extension of the rationals. In this case the components of the vector $\mathbf{v}$ are in the rational field $\mathbf{Q}$ if and only if the spectral components in the field $\mathbf{Q}^m$ satisfy

$$V_{mk}(x) = V_k(x^m) \quad (\mathrm{mod}\ p(x)).$$

The proof is similar to the previous cases.

3.5   Decimation of Shift Register Sequences

The cyclic complexity property combined with the decimation property leads immediately to the following consequence describing the feedback polynomial for a decimated shift register sequence.

□ **Theorem 3.5.1** *Let* $\mathrm{GCD}(b, n) = 1$. *If*

$$\Lambda(x) = \prod_{\ell=1}^{t}(1 - x\omega^{i_\ell})$$

is the minimal feedback polynomial to cyclically generate the vector $\mathbf{V}$, *then*

$$\Lambda'(x) = \prod_{\ell=1}^{t}(1 - x\omega^{bi_\ell})$$

is the minimal feedback polynomial for the cyclic decimation of $\mathbf{V}$ *by* b.

Proof Because $\Lambda(x)$ is the locator polynomial of $\mathbf{V}$, the inverse Fourier transform of $\mathbf{V}$ has a nonzero component v_{i_ℓ} if and only if $\Lambda(\omega^{-i_\ell}) = 0$. Let $b^{-1} = B$ (mod n). The cyclic decimation of $\mathbf{V}$ by b, denoted $\mathbf{V}'$, has an inverse Fourier transform equal to the cyclic decimation of $\mathbf{v}$ by B, denoted $\mathbf{v}'$, with components $v_i' = v_{Bi}$. Equivalently, $v_i = v'_{B^{-1}i}$. If v_i is nonzero then v'_{bi} is nonzero because

$$v'_{bi} = v_{bBi} = v_i.$$

Therefore $\Lambda'(x)$ must have its zeros at ω^{bi_ℓ} for $\ell = 1, \dots, t$.   □

With a little more work the requirement that b is relatively prime to n can be removed. Then $\Lambda'(x)$, the locator polynomial of the decimated sequence, is a product only of *unique* factors of the form $(1 - x\omega^{bi_\ell})$ and not necessarily all such unique factors are included. That is, if $bi_\ell = bi_{\ell'}$ (mod n) then the corresponding factor appears at most once in the product. Hence $\Lambda'(x)$ may have degree smaller than that of $\Lambda(x)$.

☐ **Theorem 3.5.2** *Let* $\mathrm{GCD}(b, n) = n''$, $n' = n/n''$, *and* $b' = b/n''$. *If*

$$\Lambda(x) = \prod_{\ell=1}^{t}(1 - x\omega^{i_\ell})$$

is the locator polynomial for $\mathbf{V}$ *then* $\Lambda'(x)$, *the locator polynomial for* $\mathbf{V}' = (V_{bk}, \; k = 0, \ldots, n' - 1)$, *is a product of distinct terms of the form* $(1 - x\gamma^{b'i_\ell}) = (1 - x\omega^{bi_\ell})$, *where* $\gamma = \omega^{n''}$ *and the period of* $\mathbf{V}'$ *is a divisor of* n'.

Proof The proof is similar to the proof of Theorem 3.5.1 but the decimation is not relatively prime. Because $\Lambda(x)$ is the locator polynomial of $\mathbf{V}$, the inverse Fourier transform of $\mathbf{V}$ has a nonzero component v_{i_ℓ} if and only if $\Lambda(\omega^{-i_\ell}) = 0$. Let $B'b' = 1$ (mod n'). The cyclic decimation of $\mathbf{V}$ by b, denoted $\mathbf{V}'$ has an inverse Fourier transform $\mathbf{v}'$, with components v'_i. By Theorem 3.4.3 the vector $\mathbf{v}'$ is equal to the cyclic decimation by B' of $\bar{\mathbf{v}}$, a folded version of $\mathbf{v}$. The folded vector is given by

$$\bar{v}_{i'} = \frac{1}{n''} \sum_{i''=0}^{n''-1} v_{i'+n'i''}$$

and can be nonzero only if $v_{i'+n'i''}$ is nonzero for some i''.

If $\bar{v}_{i'}$ is nonzero, then $v'_{b'i'}$ is nonzero because

$$v'_{b'i} = \bar{v}_{b'B'i}.$$

Therefore $\Lambda'(x)$ can have its zeros only at $\omega^{bi_\ell} = \gamma^{b'i_\ell}$ for $\ell = 1, \ldots, t$. ☐

3.6   A Universal Eigenvector

Does there exist a vector whose Fourier transform is proportional to itself? When the blocklength is a prime p, the vector whose components are Legendre symbols has this property. The ith component, for $i = 0, \ldots, p - 1$, is defined as $\chi_i = \chi(i)$. We shall call this vector the *Legendre vector*. Such a vector has only components equal to 0 or ± 1, so it can be regarded as a vector over any field F. We shall see in the next theorem that it is an eigenvector of the Fourier transform operator of blocklength p in F (provided a Fourier transform of blocklength p exists in F).

☐ **Theorem 3.6.1** *In any field* F, *not of characteristic* p, *the Legendre vector of prime blocklength* p *is an eigenvector of the Fourier transform operator of blocklength* p *and the corresponding eigenvalue is the Gaussian sum* θ.

Proof Let ω be an element of order p in the field F. The Fourier transform of χ, with components denoted X_k, is

$$X_k = \sum_{i=0}^{p-1} \chi_i \omega^{ik}.$$

Using the fact that $\chi_i \chi_j = \chi_{ij}$ this becomes

$$
\begin{aligned}
X_k &= \chi_k \sum_{i=0}^{p-1} \chi_{ik} \omega^{ik} \\
&= \chi_k \sum_{i=0}^{p-1} \chi_i \omega^i.
\end{aligned}
$$

The last line follows because k and p are relatively prime, so ik modulo p ranges over the same values as i modulo p. Therefore

$$X_k = \theta \chi_k,$$

where θ is the Gaussian sum. □

Although the field F is arbitrary and ω is any element in that field of order p, the Gaussian sum always takes the value $\sqrt{p}$ or $\sqrt{-p}$ (as interpreted in the field F), as stated in the following theorem.

□ **Theorem 3.6.2** *The Gaussian sum satisfies*

$$\theta^2 = p\chi(-1).$$

Proof Let

$$X_k = \sum_{i=0}^{p-1} \chi_i \omega^{ik} = \theta \chi_k$$

be the Fourier transform of χ. Because $\chi_0 = 0$ and $\chi_k = \pm 1$ for $k \neq 0$, we have

$$X_k^2 = \theta^2 - \theta^2 \delta_{k0},$$

where $\delta_{k0} = 1$ if $k = 0$ and otherwise $\delta_{k0} = 0$. The convolution theorem then allows us to express the cyclic convolution

$$(\chi * \chi)_j = \sum_{i=0}^{p-1} \chi_i \chi_{((j-i))}$$

as the inverse Fourier transform of X_k^2. Thus

$$
\begin{aligned}
(\chi * \chi)_j &= \frac{1}{p} \sum_{k=0}^{p-1} (\theta^2 - \theta^2 \delta_{k0}) \omega^{-jk} = \frac{\theta^2}{p} \sum_{k=0}^{p-1} \omega^{-jk} - \frac{\theta^2}{p} \\
&= \begin{cases} \theta^2 - \frac{\theta^2}{p}, & j = 0, \\ -\frac{\theta^2}{p}, & j \neq 0. \end{cases}
\end{aligned}
$$

The zero component of the convolution also can be computed directly as follows.

$$(\chi * \chi)_0 \;=\; \sum_{i=0}^{p-1} \chi(i)\chi(-i) = \sum_{i=1}^{p-1} \chi(i^2)\chi(-1)$$
$$=\; (p-1)\chi(-1)$$

because i^2 is always a square. Hence

$$(p-1)\chi(-1) = \theta^2 - \frac{\theta^2}{p}$$

and

$$\frac{\theta^2}{p} = \chi(-1),$$

as was to be proved.   □

The preceding ideas can be generalized to Fourier transforms whose blocklength n is a power of an odd prime p. In any field F, let ω be an element of order p^m and let

$$\theta_m = \sum_{i=0}^{p^m-1} \chi(i)\omega^i,$$

provided such an ω exists. The symbol $\chi(i)$ is defined as 0, +1, or −1 according to whether i is zero (mod p), a square (mod p^m), or a nonsquare (mod p^m).

The following theorem shows that, for $m \geq 2$, $\theta_m = 0$ in any field F. Thus

$$\sum_{i=0}^{p^m-1} \chi(i)\omega^i = 0$$

leading to a large collection of identities, including trigonometric identities, which are related to the cyclotomic polynomials. In particular the theorem implies that the polynomial

$$g(x) = \sum_{i=0}^{p^m-1} \chi(i)x^i$$

is divisible by the cyclotomic polynomial $\Phi_{p^m}(x)$.

□ **Theorem 3.6.3** *Let F be any field whose characteristic is not p, and let ω be an element of F of order p^m with $m > 1$. Then $\theta_m^2 = 0$.*

Proof It is well known that the set of positive integers less than p^m and relatively prime to p^m is cyclic under integer multiplication modulo p^m.

(We do not provide a proof of this.) Let π have order $p^m - p^{m-1}$ under multiplication of integers modulo p^m. Let

$$\chi(i) = \begin{cases} 0 & \text{if } i \text{ equals zero (mod } p), \\ 1 & \text{if } i = \pi^\ell \text{ (mod } p^m) \text{ for } \ell \text{ even,} \\ -1 & \text{if } i = \pi^\ell \text{ (mod } p^m) \text{ for } \ell \text{ odd.} \end{cases}$$

Let $\mathbf{X}$ be the Fourier transform of χ:

$$X_k = \sum_{i=0}^{p^m-1} \chi_i \omega^{ik}$$

$$= \chi_k \sum_{\ell=0}^{p^m-p^{m-1}} \chi_{\pi^\ell k} \omega^{\pi^\ell k}.$$

For $k \neq 0$ (mod p), this can be written

$$X_{\pi^r} = \chi_{\pi^r} \sum_{\ell=0}^{p^m-p^{m-1}} \chi_{\pi^{\ell+r}} \omega^{\pi^{\ell+r}},$$

where $\pi^r = k$. Because ℓ ranges over a complete cycle, the sum is independent of r. Therefore

$$X_k = \chi_k \sum_{i=0}^{p^m-1} \chi_i \omega^i = \chi_k \theta_m$$

if $\mathrm{GCD}(k, p) = 1$.

For the remaining p^{m-1} values of k, first suppose $\mathrm{GCD}(p, k) = p$. There are $p^{m-1} - p^{m-2}$ such values of k. Because $\chi_{i+p} = \chi_i$, we can write

$$X_{pk'} = p \sum_{i=0}^{p^{m-1}-1} \chi_i \gamma^{ik'},$$

where $\gamma = \omega^p$. The sum is the same form seen earlier when k is not divisible by p, so

$$X_{pk'} = \pm p\theta_{m-1}.$$

More generally, if $\mathrm{GCD}(p, k) = p^s$,

$$X_{p^s k'} = \pm p^s \theta_{m-s}.$$

There are $p^{m-s} - p^{m-s-1}$ such values of k.

We will sum X_k^2 to get $(\chi * \chi)_0$:

$$(\chi * \chi)_0 = \frac{1}{p^m} \sum_{k=0}^{p^m-1} X_k^2$$

$$= \frac{1}{p^m} \sum_{\ell=0}^{m-1} p^{2\ell} \theta_{m-\ell}^2 (p^{m-\ell} - p^{m-\ell-1})$$

$$= \frac{1}{p^m} \sum_{i=1}^{m} \theta_i^2 p^{2(m-i)} (p^i - p^{i-1}).$$

On the other hand,

$$(\chi * \chi)_0 = \sum_{i=0}^{p^m-1} \chi(i) \, \chi(-i) = \sum_{i=0}^{p^m-1} \chi(i^2) \, \chi(-1)$$

$$= (p^m - p^{m-1}) \, \chi(-1)$$

$$= \frac{1}{p^m} \theta_1^2 p^{2(m-1)} (p - 1).$$

Equating the two expressions for $(\chi * \chi)_0$ and canceling the terms in θ_1^2 gives $\sum_{i=2}^{m} \theta_i^2 p^{2(m-i)} (p^i - p^{i-1}) = 0$. This implies that θ_m^2 equals zero for $m \geq 2$ because such an expression holds for each m larger than 1.   □

3.7   Bounds on the Weight of Sequences

We have seen that the weight of a sequence is equal to the cyclic complexity of its Fourier transform. This property immediately leads to bounds on the weight of a sequence based on properties of the spectrum.

□ **Theorem 3.7.1 (Bose–Chaudhuri–Hocquenghem (BCH) Bound)**
Suppose that the field F contains an element of order n. The only vector of blocklength n in the field F of weight $d-1$ or less that has $d-1$ (cyclically) consecutive components of its spectrum equal to zero is the all-zero vector.

Proof Let $\mathbf{v}$ be a vector of weight $d-1$ or less. The cyclic complexity of $\mathbf{V}$ is equal to the weight of $\mathbf{v}$. Therefore

$$V_k = -\sum_{j=1}^{d-1} \Lambda_j V_{((k-j))}.$$

If $\mathbf{V}$ has $d-1$ consecutive components equal to zero, the recursion implies that $\mathbf{V}$ is everywhere zero and that $\mathbf{v}$ must be the all-zero vector.   □

Two other bounds generalize the BCH bound. The next theorem says that the consecutive $d-1$ spectral zeros of the BCH bound can be replaced by a pattern of s uniformly spaced subblocks each of $d-s$ consecutive spectral zeros. The modulation property of the Fourier transform implies that any pattern of spectral zeros can be cyclically shifted without changing the linear complexity. Therefore it is enough to choose the pattern of spectral zeros beginning at index $k = 0$.

□ **Theorem 3.7.2 (Hartmann–Tzeng Bound)** *Suppose the field F contains an element of order n and let* $\text{GCD}(n, b) = 1$. *The only vector in the field F of blocklength n and of weight $d - 1$ or less whose spectral components V_k equal zero for $k = j_1 + j_2 b$, where $j_1 = 0, \ldots, d - 1 - s$ and $j_2 = 0, \ldots, s - 1$, is the all-zero vector.*

Proof The idea of the proof is to construct a new spectrum, possibly in an extension field, whose cyclic complexity is at least as large and to which the BCH bound can be applied. Each step of the proof is based on elementary properties of the Fourier transform.

Let **V** be the spectrum under consideration and **v** its inverse Fourier transform. By the modulation property of Fourier transforms, we know that if **V** is cyclically shifted the components of the inverse Fourier transform **v** that were nonzero remain nonzero and the components that were zero remain zero. Adding two components that are each zero cannot create a nonzero component. Therefore, the weight of the inverse Fourier transform cannot increase under linear combinations of cyclic shifts of **V**. Consequently, the cyclic complexity of **V** cannot increase under linear combinations of its cyclic shifts.

Let $\mathbf{V}^{(\ell)}$ be the cyclic shift of **V** given by

$$V_k^{(\ell)} = V_{((k+\ell))}, \qquad \ell = 0, \ldots, d - s - 1.$$

Each $\mathbf{V}^{(\ell)}$ has a zero at $k = 0$ because $V_0^{(\ell)} = V_\ell$ and $V_\ell = 0$ for $\ell = 0, \ldots, d - 1 - s$. In the same way, $V_{jb}^{(\ell)} = 0$ for $\ell = 0, \ldots, d - 1 - s$ and $j = 0, \ldots, s - 1$. The linear combination

$$\mathbf{V}' = \sum_{\ell=0}^{d-1-s} a_\ell \mathbf{V}^{(\ell)}$$

is zero in each component where every $\mathbf{V}^{(\ell)}$ equals zero. Thus

$$V_{jb}' = 0 \qquad \text{for } j = 0, \ldots, s - 1$$

and V_k' may be zero in other components as well, depending on the choice of a_ℓ.

The set of such vectors $\mathbf{V}^{(\ell)}$ is linearly independent because the set of their transforms $(v_i \omega^{i\ell})$ is linearly independent, so we may choose the $d - s$ arbitrary coefficients a_ℓ to obtain the $d - s - 1$ additional zeros:

$$C_{jb}' = 0 \quad \text{for } j = s, \ldots, d - 2.$$

Now **C**$'$ has spectral zeros at $k = jb$ for $j = 0, \ldots, d - 2$.

Finally, because $\text{GCD}(b, n) = 1$, the components of the spectrum of **C**$'$ can be permuted by cyclic decimation, beginning with component C_0' and taking every bth component of **C**$'$ cyclically. This permutation of the spectrum corresponds to a permutation of the components of the inverse Fourier transform; hence the linear complexity is not affected, but now the $d - 1$ zero components of the spectrum are consecutive components. The BCH bound now implies that

$$w(\mathbf{c}') \geq d$$

for the vector $\mathbf{c}'$ and, by construction of $\mathbf{c}'$, the weight of $\mathbf{c}$ is not smaller.
□

□ **Corollary 3.7.3 (Roos Bound)** *Let n be a factor of $q^m - 1$ for some m. Let $\mathrm{GCD}(n, b) = 1$. The only vector in $GF(q)^n$ of weight $r + s - 1$ or less whose spectral components equal zero for $k = j_1 + j_2 b$, where $j_1 = 0, \ldots, d - 1 - s$, and for at least s values of j_2 in the range $0, \ldots, d - 2$ is the all-zero vector.* □

Proof The proof is the same as the proof of Theorem 3.7.2. The linear combination is chosen to make $C'_{jb} = 0$ for the $d - s - 1$ missing values of j, which need not be restricted to a particular pattern of missing values. □

3.8 The Gleason–Prange Theorem

Elementary properties of the discrete Fourier transform were developed in Section 3.4. In this section, which subsequent chapters do not depend on, we shall develop another general property of the Fourier transform, but one that is too narrow in its scope and usefulness to be included in Section 3.4. It deals with a rather strange permutation of a sequence whose length is a prime. The applications of this theorem are few, but the theorem is included in keeping with our plan of providing a complete catalog of general properties of the discrete Fourier transform. Perhaps new applications will be found in the future.

Let p be a prime. We will consider a vector $\mathbf{c}$ of blocklength $p + 1$,

$$\mathbf{c} = (c_0, c_1, \ldots, c_{p-1}, c_\infty)$$

over any field F with components indexed by the elements of $GF(p) \cup \{\infty\}$. The component c_∞ has a distinctive index because it will be treated differently than the other components. We emphasize that F and $GF(p)$ are distinct fields; the components of the vector $\mathbf{c}$ are elements of F and the indices of $\mathbf{c}$ are elements of $GF(p)$.

For every nonzero element i of $GF(p)$, the elements $\frac{1}{i}$ and $-\frac{1}{i}$ are defined. The special element ∞ is appended to $GF(p)$ so that $-\frac{1}{0}$ can be defined as ∞. Then every element of $GF(p) \cup \{\infty\}$ has a negative inverse defined with the understanding that $-\frac{1}{0} = \infty$ and $-\frac{1}{\infty} = 0$.

The Fourier transform of $\mathbf{c}$ is defined on the first p components:

$$C_k = \sum_{i=0}^{p-1} w^{ik} c_i.$$

The Gleason–Prange permutation of $\mathbf{c}$ is defined as the vector

$$\mathbf{d} = (d_0, d_1, \ldots, d_{p-1}, d_\infty),$$

where

$$
\begin{aligned}
d_i &= \chi(-1/i)c_{-1/i}, & i = 1, \ldots, p-1, \\
d_0 &= \chi(-1)c_\infty, \\
d_\infty &= c_0,
\end{aligned}
$$

and $\chi(i)$ is the Legendre symbol.

□ **Theorem 3.8.1 (Gleason–Prange)** *Let p be a prime. Suppose that over the field F, the vector*

$$\mathbf{c} = (c_0, c_1, \ldots, c_{p-1}, c_\infty)$$

satisfies

i) $C_k = 0$ *if k is a nonzero square,*

ii) $c_\infty = -\dfrac{\theta}{p} \sum_{i=0}^{p-1} c_i,$

where θ is the Gaussian sum. Then the Gleason–Prange permutation of $\mathbf{c}$ satisfies these same two conditions.

Proof The proof consists of verifying in turn that each of the two conditions holds.

i) $D_k = 0$ if k is a nonzero square

ii) $d_\infty = -\dfrac{\theta}{p} \sum_{i=0}^{p-1} d_i.$

Proof of i). Let $\mathbf{C}$ be the Fourier transform of $\mathbf{c}$ and $\mathbf{D}$ be the Fourier transform of $\mathbf{d}$. By definition C_k is zero if k is a nonzero square. To get $\mathbf{D}$ from $\mathbf{C}$, first compute the inverse Fourier transform of $\mathbf{C}$ to get $\mathbf{c}$; next permute components of $\mathbf{c}$ to get $\mathbf{d}$; then take the direct Fourier transform of $\mathbf{d}$ to get $\mathbf{D}$. We will show that this sequence of operations preserves the locations of the zeros defining the spectrum.

The inverse Fourier transform can be written

$$c_i = \frac{1}{p}\left[C_0 + \sum_{k=1}^{p-1} w^{-ik} C_k\right] = \frac{1}{p}\left[\frac{-p}{\theta}c_\infty + \sum_{k=1}^{p-1} w^{-ik} C_k\right].$$

Under the Gleason–Prange permutation,

$$d_i = \chi\left(-\frac{1}{i}\right)c_{-1/i} = \frac{1}{p}\chi\left(-\frac{1}{i}\right)\left[\frac{-p}{\theta}c_\infty + \sum_{k=1}^{p-1} w^{\frac{1}{i}k} C_k\right]$$

and

$$d_0 = \chi(-1)c_\infty.$$

These expressions can be substituted into the expression for the Fourier transform of **d** as follows:

$$
\begin{aligned}
D_j &= d_0 + \sum_{i=1}^{p-1} w^{ij} d_i \\
&= \chi(-1)c_\infty + \sum_{i=1}^{p-1} w^{ij}\frac{1}{p}\chi\left(-\frac{1}{i}\right)\left[\frac{-p}{\theta}c_\infty + \sum_{k=1}^{p-1} w^{\frac{1}{i}k} C_k\right] \\
&= \chi(-1)c_\infty\left[1 - \theta^{-1}\sum_{i=1}^{p-1}\chi(i)w^{ij}\right] + \frac{1}{p}\sum_{i=1}^{p-1} w^{ij}\chi\left(-\frac{1}{i}\right)\sum_{k=1}^{p-1} w^{\frac{1}{i}k} C_k,
\end{aligned}
$$

where we have used $\chi\left(-\frac{1}{i}\right) = \chi(-1)\chi(i)$ in forming the first term. But j is relatively prime to p so the sum in the first term reduces to $\chi(j)\theta$. Therefore

$$D_j = \chi(-1)c_\infty[1 - \chi(j)\theta^{-1}\theta] + \frac{1}{p}\chi(-1)\sum_{i=1}^{p-1} w^{ij}\sum_{k=1}^{p-1} w^{\frac{1}{i}k}\chi\left(\frac{1}{i}k\right)\chi(k)C_k.$$

The term $\chi(k)$ in the second term can be replaced by -1 because $C_k = 0$ whenever $\chi(k) \ne -1$. Thus

$$D_j = \chi(-1)\left[c_\infty(1 - \chi(j)) - \frac{1}{p}\sum_{i=1}^{p-1} w^{ij}\sum_{k=1}^{p-1} w^{\frac{1}{i}k}\chi\left(\frac{1}{i}k\right)C_k\right].$$

The first term is zero if j is a nonzero square, so we ignore this term henceforth by restricting attention only to those j that are nonzero squares. Next observe that the indices are elements of the field $GF(p)$, so the nonzero indices can be written as powers of a primitive element π. Let

$$i = \pi^r, \quad k = \pi^t, \quad j = \pi^{-s}.$$

Replacing i and k by r and t these are permutations (called Rader permutations in signal processing), so the sums are unaffected. Consequently,

$$
\begin{aligned}
D_{\pi^{-s}} &= -\frac{1}{p}\chi(-1)\sum_{r=0}^{p-2}\omega^{\pi^r\pi^{-s}}\sum_{t=0}^{p-2}\omega^{\pi^{-r}\pi^t}\chi(\pi^{-r}\pi^t)C_{\pi^t} \\
&= -\frac{1}{p}\chi(-1)\sum_{r=0}^{p-2}\omega^{\pi^{-s+r}}\sum_{t=0}^{p-2}(-1)^{-r+t}\omega^{\pi^{-r+t}}C_{\pi^t},
\end{aligned}
$$

where $\chi(\pi^{t-r})$ has been replaced by its equivalent $(-1)^{t-r}$. This now has the form of a double cyclic convolution:

$$
D'_{-s} = \sum_{r=0}^{p-2} g_{r-s}\sum_{t=0}^{p-2} g'_{t-r}C'_t,
$$

where

$$
\begin{aligned}
C'_t &= C_{\pi^t}, \\
D'_s &= \left[-\frac{1}{p}\chi(-1)\right]^{-1}D_{\pi^s}, \\
g_r &= \omega^{\pi^{-r}}, \\
g'_r &= (-1)^r\omega^{\pi^{-r}}.
\end{aligned}
$$

If k is a nonzero square, C_k is zero. Hence if t is even, C'_t is zero. In polynomial notation, the double convolution is written

$$
D'\left(\frac{1}{x}\right) = g(x)g'(x)C'(x) \qquad (\bmod\ x^{p-1}-1),
$$

where

$$
\begin{aligned}
g(x) &= \sum_{r=0}^{p-2}\omega^{\pi^{-r}}x^r, \\
g'(x) &= \sum_{r=0}^{p-2}(-1)^r\omega^{\pi^{-r}}x^r.
\end{aligned}
$$

Because these differ only in the sign of the odd-indexed terms, the product $g(x)g'(x)$ has only even-indexed coefficients nonzero. By definition, $C'(x)$ has only odd-indexed coefficients nonzero. Therefore $g(x)g'(x)C'(x)$ has all even-indexed coefficients equal to zero. Hence $D'_s = 0$ if s is even, so $D_j = 0$ if j is a nonzero square. This completes the proof of part i.

Proof of ii).

$$
\begin{aligned}
\sum_{i=0}^{p-1} d_i &= \chi(-1)c_\infty + \sum_{i=1}^{p-1}\chi(-1/i)c_{-1/i} \\
&= \chi(-1)c_\infty + \sum_{i=1}^{p-1}\chi(i)c_i.
\end{aligned}
$$

Replacing c_i by the expression of the inverse Fourier transform gives

$$\sum_{i=1}^{p-1} \chi(i) c_i = \frac{1}{p} \sum_{i=1}^{p-1} \sum_{k=1}^{p-1} \chi(i) [\omega^{-ik} C_k + C_0].$$

Because $\chi(i)$ equals 1 for $(p-1)/2$ values of i and equals -1 for $(p-1)/2$ values of i the term in C_0 is zero. Then, using the definition of the Gaussian sum, we have

$$\sum_{i=0}^{p-1} d_i = \chi(-1) c_\infty + \frac{1}{p} \chi(-1) \sum_{k=1}^{p-1} C_k \chi(k) \theta.$$

But $C_k = 0$ whenever $\chi(k) \neq -1$. Therefore

$$\begin{aligned}
\sum_{i=0}^{p-1} d_i &= \chi(-1) \left[c_\infty - \theta \frac{1}{p} \sum_{k=1}^{p-1} C_k \right] \\
&= -\chi(-1) \theta \frac{1}{p} \sum_{k=0}^{p-1} C_k
\end{aligned}$$

because $c_\infty = -(\theta/p) C_0$. Therefore

$$\sum_{i=0}^{p-1} d_i = -\chi(-1) \theta c_0 = -\chi(-1) \theta d_\infty.$$

Consequently, because $\theta^2 \chi(-1) = p$,

$$\sum_{i=0}^{p-1} d_i = -\frac{p}{\theta} d_\infty,$$

as was to be proved. $\square$

4

Cyclic Codes and Related Codes

Just as the FIR filter, which is closely related to convolutions, Fourier transforms, and Toeplitz matrices, is a central topic of digital signal processing, so the cyclic code is a central topic of error-control codes. A cyclic code is simply an error-control code that satisfies a certain convolution property in the time domain or a certain spectral property in the frequency domain. We will develop the theory of cyclic codes and related codes by working as much as possible within the framework of digital signal processing.

4.1 Theory of Reed–Solomon Codes

Our task is to construct a set of vectors of blocklength n with the property that for any vector of the set, after the values of as many as t arbitrary components of that vector are replaced by any other values, it will always be possible to recover the original vector. This means that a data block can be represented by a unique vector of blocklength n which will be called a *codeword*. The codeword is transmitted through a channel that is allowed to make as many as t component errors and the correct codeword will still be evident. The set of codewords then is called a t-error-correcting code.

We will design such a set of vectors with the property that every two vectors in the set differ in at least $2t + 1$ components. Changing one vector in t places gives a new vector that differs from the original vector in t places and differs from every other vector in at least $t+1$ places. Hence the correct codeword is the one that differs from the received word in the fewest places. This is said to be the codeword at the minimum Hamming distance from the received word. In general, the *Hamming distance* between two vectors of length n is defined as the number of places in which they differ.

To define the set, recall the BCH bound which says that $\mathbf{c}$ has weight at least $2t + 1$ if $\mathbf{C}$ has $2t$ consecutive components equal to zero. Hence $\mathbf{c} - \mathbf{c}'$ has weight at least $2t + 1$ if $\mathbf{C} - \mathbf{C}'$ has $2t$ consecutive components equal to zero. Hence $\mathbf{c}$ and $\mathbf{c}'$ differ in at least $2t + 1$ places if $\mathbf{C}$ and $\mathbf{C}'$ have the same $2t$ consecutive components equal to zero. A Reed–Solomon code is the set of all vectors with a fixed set of $2t$ consecutive components equal to zero; any two such vectors must differ in at least $2t + 1$ places.

The following definition chooses to set $C_k = 0$ for $k = n - 2t, n - 2t + 1,$ $\ldots, n - 1$. Any other set of $2t$ consecutive components would serve just as well.

□ **Definition 4.1.1** *Let F contain an element of order n. The $(n, n - 2t)$ Reed–Solomon code of blocklength n with symbols in F is the set of all vectors $\mathbf{c}$ whose spectrum (in F) satisfies $C_k = 0$ for $k = -1, -2, \ldots, -2t$. This is described briefly as an $(n, n - 2t)$ Reed–Solomon code over F.* □

We emphasize that the spectrum of a Reed–Solomon codeword is in the same field as the code.

One way to find the Reed–Solomon codewords is to encode in the frequency domain. This means setting $C_k = 0$ for $k = -1, \ldots, -2t$ and setting the remaining $n - 2t$ components of the transform equal to the $n - 2t$ data symbols given by $d_0, \ldots, d_{n-2t-1}$. That is,

$$C_k = \begin{cases} d_k, & k = 0, \ldots, n - 2t - 1, \\ 0, & k = n - 2t, \ldots, n - 1. \end{cases}$$

(Recall that C_n and C_0 are the same because $n = 0 \pmod{n}$.)

An inverse Fourier transform produces the codeword $\mathbf{c}$. The number of data symbols encoded equals $n - 2t$ and there are $2t$ extra symbols in the codeword to correct t errors. A Reed–Solomon code always uses two overhead symbols for every error to be corrected.

Using the Fourier transform is not the only way to map the $n - 2t$ data symbols into codewords — others may yield a simpler implementation — but the frequency-domain encoder is the most convenient method to start with because it exhibits very explicitly the notion that the codewords are those words with the same set of $2t$ zeros in the transform domain.

An alternative encoder in the time domain works as follows. The $n - 2t$ data symbols are expressed as a polynomial

$$d(x) = d_{n-2t-1}x^{n-2t-1} + d_{n-2t-2}x^{n-2t-2} + \ldots + d_1 x + d_0,$$

where $d_0, d_1, \ldots, d_{n-2t-1}$ are the $n - 2t$ data symbols in the field F. The n codeword symbols are given by the polynomial product

$$c(x) = g(x)d(x),$$

where $g(x)$ is a fixed polynomial called the *generator polynomial*. This is nothing more than a FIR filter. The generator polynomial is the polynomial

$$g(x) = (x - \omega^{-1})(x - \omega^{-2}) \ldots (x - \omega^{-2t}).$$

The kth component of the Fourier transform of the codeword is formally the same as evaluating the polynomial $c(x)$ at ω^k. That is,

$$C_k = \sum_{i=0}^{n-1} c_i \omega^{ik} = c(\omega^k) = g(\omega^k)d(\omega^k).$$

The definition of $g(x)$ ensures that $C_k = 0$ for $k = -1, \ldots, -2t$.

Both of these methods of encoding have the property that the data symbols do not appear explicitly in the codeword. Another method of encoding, known as *systematic encoding*, leaves the data symbols unchanged and exhibited in the first $n - 2t$ components of the codeword. Multiplication of $d(x)$ by x^{2t} will move the components of $d(x)$ left $2t$ places. Thus we can write

$$c(x) = x^{2t}d(x) + r(x),$$

where $r(x)$ is a polynomial of degree less than $2t$ and is appended to make the spectrum be a legitimate codeword spectrum. The spectrum will be right if $c(x)$ is a multiple of $g(x)$, and this will be so if $r(x)$ is chosen as the negative of the remainder when $x^{2t}d(x)$ is divided by $g(x)$. Thus

$$c(x) = x^{2t}d(x) - R_{g(x)}[x^{2t}d(x)]$$

gives a systematic encoder, where the operator $R_{g(x)}$ takes the remainder under division by $g(x)$. To see that $c(x)$ is a codeword, simply notice that $c(x)$ has no remainder under division by $g(x)$.

The Reed–Solomon decoder finds the codeword that agrees most closely with the received word. It does not depend on how the codewords are used to store data symbols except for the final step of reading the data out of the corrected codeword.

For the moment, we will imagine that we can find the closest codeword by exhaustively searching for it, comparing each codeword in turn to the received word. In Chapter 8, we will discuss some computational procedures that yield practical algorithms for finding the correct codeword directly.

The Reed–Solomon code could be defined just as well by constraining any other $2t$ consecutive spectral components to be equal to zero. Thus we could define the alternative code by

$$C_k = 0, \qquad k = k_0 - 1, \ldots, k_0 - 2t.$$

The modulation/translation property of Fourier transforms implies that this is the same as the earlier definition except that the ith component of every codeword is multiplied by ω^{-ik_0}, which does not alter the number of components in which two codewords are different.

The BCH bound proves that t errors in any codeword of a Reed–Solomon code can always be corrected, because it proves that every pair of codewords in the code differ in at least $2t+1$ places. Because this is true, changing any t components of a codeword will produce a word that is different from the correct codeword in t places and is different from every other codeword in at least $t + 1$ places. If at most t errors occur, then choosing the codeword that differs from the noisy received word in the fewest number of places will recover the correct codeword. If each symbol is more likely to be correct than to be in error, then choosing the codeword that differs from the noisy received word in the fewest places will recover the most likely codeword and will minimize the probability of decoding error.

The following theorem follows from the discussion above, but we also provide a more direct proof based on the fundamental theorem of algebra.

☐ **Theorem 4.1.2** *Any two codewords in an* $(n, n - 2t)$ *Reed–Solomon code differ in at least* $2t + 1$ *places.*

Proof The difference between two codewords must also have a spectrum that is zero for $k = n - 2t, \ldots, n - 1$ and so itself is a codeword. We only need to prove that no codeword has fewer than $2t + 1$ nonzero components unless it is zero in every component. Let

$$C(y) = \sum_{k=0}^{n-2t-1} C_k y^k.$$

This is a polynomial of degree at most $n - 2t - 1$, so by the fundamental theorem of algebra it has at most $n - 2t - 1$ zeros. Thus the inverse Fourier transform

$$c_i = \frac{1}{n} C(\omega^{-i})$$

can be zero in at most $n - 2t - 1$ places and so it is nonzero in at least $2t + 1$ places. This completes the proof.   ☐

4.2   Reed–Solomon Codes in Infinite Fields

The proof that the Reed–Solomon code corrects t errors is valid in any field. All that were used in the proof were the fundamental theorem of algebra and the properties of the Fourier transform. These hold in any field. We can construct Reed–Solomon codes in the complex field or extensions of the rational field so that the code symbols naturally correspond to the channel input and output but the encoder and decoder must contend with precision problems in the computations. Problems of precision are why such codes have not seen applications; perhaps they will play some future role.

4.2.1   REED–SOLOMON CODES IN THE COMPLEX FIELD

An encoder for a Reed–Solomon code in the complex field maps a data vector of $n - 2t$ complex numbers into a vector of n complex numbers, a decoder for that code maps a noisy version of that codeword having up to t erroneous components back to the original codeword and then to the $n - 2t$ complex data symbols. Loosely, we may say that a Reed–Solomon code in the complex field is a signal processing technique to protect a complex vector against t noise impulses.

One might wish to use an error-control code over the complex field because computational devices that do real or complex arithmetic are widely available. To achieve a single processor that does both digital signal processing and error control, one may wish to do the error control computations with real arithmetic. Another possible reason for using the complex field is that Reed–Solomon codes of every blocklength exist in the complex field.

An important consideration in an infinite field is that of roundoff error. Whereas in a finite field it is quite specific to say how many components are in error, in the real field or the complex field there may be some minor error in every component of the received signal. An error-control code can be used to correct up to t major errors, which may be due to burst noise or impulsive noise in the channel. The correction will be successful even if all the other components have minor errors. However, to date there has been no theoretical work quantifying how big the minor errors and computational noise can be before the decoding algorithms, such as those discussed in Chapter 8, break down. Problems concerned with roundoff error and division by small numbers have not been fully studied. This topic awaits further development.

There is another side to this story. It may be that the channel is a discrete channel containing a demodulator that forms an estimate of one of a finite number of symbols. Data symbols can be fixed-point numbers of finite precision. In a systematic code over the complex field, even though the wordlength of data symbols is finite, the wordlength of parity-check symbols will not be finite.

4.2.2 REED–SOLOMON CODES IN THE EXTENDED RATIONALS

One can also form Reed–Solomon codes in $\mathbf{Q}^m$, the extension of the rationals. In contrast to Reed–Solomon codes in $\mathbf{C}$, Reed–Solomon codes in $\mathbf{Q}^m$ can be formed in which all symbols are integers. Such a code can be used with a discrete channel such as a digital memory. An advantage is that the arithmetic used by the encoder and decoder is conventional integer arithmetic if the demodulator rounds the received symbols to the nearest integer. There is no need to round off in the encoder. A disadvantage is that the parity-check symbols will require longer integer wordlength than the data symbols. If this is an issue, the wordlength of the parity-check symbols can be truncated, thereby making an occasional error in the encoder; the decoder will use the structure of the code to correct such intentional errors but the number of allowed channel errors will be reduced.

We will describe a Reed–Solomon code in $\mathbf{Q}^m$ as an example, representing the elements of $\mathbf{Q}^m$ as polynomials in y of degree $m-1$. Codewords are polynomials in x whose symbols are polynomials in y. The generator polynomial is

$$g(x) = (x - \omega)(x - \omega^2)(x - \omega^3) \ldots (x - \omega^{2t}).$$

If m is a power of 2 in $\mathbf{Q}^m$, we have the Fourier transform

$$C_k = \sum_{i=0}^{n-1} y^{ik} c_i, \qquad k = 0, \ldots, n - 1,$$

where ω has been chosen as the field element y, and all arithmetic is poly-nomial arithmetic modulo the cyclotomic polynomial $p(y)$. Consequently, the generator polynomial $g(x)$ is

$$g(x) = (x - y)(x - y^2)(x - y^3) \ldots (x - y^{2t}) \quad (\bmod\ p(y)),$$

which has only integer coefficients and is actually a polynomial in two variables. To emphasize this, we may write $g(x, y)$.

Specifically, let $m = 64$ and let $p(y) = y^{32} + 1$. Then the Fourier trans-form has blocklength 64. Codewords are vectors of length 64 of elements of $\mathbf{Q}^{32}$. That is, codewords are vectors of length 64 of polynomials of degree 31. They can be viewed as two-dimensional arrays or as two-dimensional polynomials and written as

$$c(x, y) = \sum_{i=0}^{63} \sum_{j=0}^{31} c_{ij} x^i y^j,$$

where the c_{ij} are rational numbers in general but can be restricted to integers.

To encode nonsystematically, let the data be represented as

$$d(x, y) = \sum_{i=0}^{63-2t} \sum_{j=0}^{31} d_{ij} x^i y^j.$$

Then

$$c(x, y) = g(x, y) d(x, y) \quad (\bmod\ y^{32} + 1).$$

If the data components d_{ij} take only integer values, then the codeword components c_{ij} also take only integer values.

The received word $v(x, y)$ is a vector of length 64 of elements of $\mathbf{Q}^{32}$. That is, the received word is a vector of length 64 of polynomials of degree 31. Any t polynomials can be in error and the correct codeword will be recoverable by the structure of the code.

4.3   Reed–Solomon Codes in Finite Fields

The Reed–Solomon codes have been most popular in the Galois fields of characteristic 2, $GF(2^m)$. Each symbol can be used to represent m bits, so there is a nice match between the code and data represented in binary form.

In $GF(8)$ there is an element of order 7 (in fact, every nonzero element of $GF(8)$ has order 7). Choose $p(x) = x^3 + x + 1$ and write the field in exponential form as

$$
\begin{aligned}
\alpha &= x \\
\alpha^2 &= x^2 \\
\alpha^3 &= x + 1 \\
\alpha^4 &= x^2 + x \\
\alpha^5 &= x^2 + x + 1 \\
\alpha^6 &= x^2 + 1 \\
\alpha^7 &= 1.
\end{aligned}
$$

To construct the code choose $\omega = \alpha = x$. Then

$$
\begin{aligned}
g(y) &= (y - \omega^1)(y - \omega^2) \\
&= y^2 + \alpha^4 y + \alpha^3.
\end{aligned}
$$

This completes the design of the code. If desired to emphasize the fact that the field elements are represented as three-bit numbers, we may write

$$g(y) = (001)y^2 + (110)y + (011).$$

Figure 4.1 shows some of the codewords of the (7,5) Reed–Solomon code. Altogether, the full list contains $8^5 = 2^{15}$ codewords.

An alternative (7,5) Reed–Solomon code over $GF(8)$ uses two other consecutive powers of ω as zeros of $g(y)$. For example,

$$g(y) = (y - \omega^3)(y - \omega^4)$$

gives a code with the same performance, but not the same code.

In $GF(256)$ there is a Fourier transform of blocklength 255 so there is a t-error-correcting Reed–Solomon code over $GF(256)$ with blocklength 255 and $255 - 2t$ data symbols. Code symbols and data symbols are eight-bit bytes. With spectral zeros at $k = 1, \ldots, 2t$, the generator polynomial is

$$g(x) = (x - \omega)(x - \omega^2) \ldots (x - \omega^{2t}).$$

All the arithmetic of the encoding and decoding operations takes place in $GF(256)$.

Because the blocklength n of a Fourier transform over $GF(2^m)$ must be a divisor of $2^m - 1$, so also must the blocklength of the code. Such a blocklength may be inconvenient in an application. There is a way to extend a Reed–Solomon code by appending an overall parity-check symbol, which

$$
\begin{array}{ccccccc}
0 & 0 & 0 & 0 & 0 & 0 & 0 \\
0 & 0 & 0 & 0 & 1 & 6 & 3 \\
0 & 0 & 0 & 0 & 2 & 7 & 6 \\
0 & 0 & 0 & 0 & 3 & 1 & 5 \\
& & & \vdots & & & \\
0 & 0 & 0 & 1 & 0 & 1 & 1 \\
0 & 0 & 0 & 1 & 1 & 7 & 2 \\
0 & 0 & 0 & 1 & 2 & 6 & 7 \\
0 & 0 & 0 & 1 & 3 & 0 & 4 \\
& & & \vdots & & & \\
0 & 0 & 0 & 7 & 0 & 7 & 7 \\
0 & 0 & 0 & 7 & 1 & 1 & 4 \\
0 & 0 & 0 & 7 & 2 & 0 & 1 \\
0 & 0 & 0 & 7 & 3 & 6 & 2 \\
& & & \vdots & & & \\
0 & 0 & 1 & 0 & 0 & 7 & 3 \\
0 & 0 & 1 & 0 & 1 & 1 & 0 \\
0 & 0 & 1 & 0 & 2 & 0 & 5 \\
0 & 0 & 1 & 0 & 3 & 6 & 6 \\
& & & \vdots & & & \\
\end{array}
$$

FIGURE 4.1. The (7,5) Reed–Solomon code.

increases the minimum Hamming distance by one. The blocklength of the Fourier transform is still $2^m - 1$, so this is the Fourier transform that the decoder must use. The reason for extending the code is that a power of 2 blocklength may be a better fit to some applications. For example, a (256,224) Reed–Solomon code would be preferable to a (255,223) Reed–Solomon code if the codeword were transferred in blocks of 256 eight-bit bytes. The additional complexity in the encoder and decoder for the extra byte is completely insignificant.

The construction of the extended encoder is as follows. Begin with a Reed–Solomon code of minimum distance $2t$ having $2t - 1$ spectral zeros at indices $k = k_0, k_0 + 1, \ldots, k_0 + 2t - 2$. This code has $n - 2t + 1$ data symbols. Let

$$
C_\infty = C_{k_0+2t-1} = \sum_{i=0}^{n-1} c_i \omega^{(k_0+2t-1)i}.
$$

(If desired k_0 can be chosen so that $k_0 + 2t - 1 = 0$.) The codeword of the extended code is

$$
\mathbf{c} = (c_0, c_1, \ldots, c_{n-1}, c_\infty).
$$

As can be seen in the decoders of Chapter 8, it is trivial to generate synthetically an extra zero in the spectrum when needed for decoding by subtracting c_∞ from the actual value of that spectral component.

4.4 Radix-2 Reed–Solomon Codes

Some prime fields admit a Fourier transform whose blocklength n is a power of 2. In any of these fields, a Reed–Solomon code exists with a power of 2 blocklength and in that field the decoder can use a radix-2 fast Fourier transform (FFT). For example, in $GF(769)$ there is a Fourier transform of blocklength 256, so we can form Reed–Solomon codes of blocklength 256 whose decoders use a 256-point FFT in $GF(769)$. The disadvantage is that the symbol alphabet is no longer a power of 2. Each symbol holds the equivalent of 9.6 bits. When the data are provided in binary form, then either the symbols must be repacked into the symbols of $GF(769)$ or else only 9 bits of the 9.6 bits can be used for data.

One can also form a Reed–Solomon code in an extension field such as $GF(769^2)$. Then each symbol is a pair of 769-ary numbers. The Reed–Solomon code over $GF(p^2)$ has little advantage over using two Reed–Solomon codes over $GF(p)$ except that the blocklength can be any divisor of $p^2 - 1$.

For example, the Mersenne field $GF(2^{13} - 1)$ has no element whose order is a power of 2, but $GF((2^{13} - 1)^2)$ does have such an element. In $GF((2^{13} - 1)^2)$ we can form a Reed–Solomon code of any blocklength that divides 2^{14}. Code symbols are elements of $GF((2^{13} - 1)^2)$ represented by pairs of 13-bit numbers. The decoding computations treat the numbers as integers modulo $2^{13} - 1$ so the number consisting of 13 ones is equal to the number consisting of 13 zeros. What this amounts to is that the decoder will be blind to the difference between the all-zeros number and the all-ones number. If there is an error in such a symbol, the decoder will arbitrarily replace the erroneous symbol by one of these two numbers, say the all-zeros number. The Reed–Solomon code in $GF((2^{13} - 1)^2)$ may be attractive if the all-ones number is prohibited in the data.

Another technique that works well if the prime p is slightly larger than a power of 2 is to make deliberate errors by rounding down any numbers that exceed the chosen wordlength. For example, one can form Reed–Solomon codes in the Fermat field $GF(2^{16} + 1)$ with blocklength any divisor of 2^{16}. This code can be used with 16-bit numbers as follows. Restrict each data symbol to 16 bits and use a systematic encoder. Parity-check symbols will almost always fit within a 16-bit number, but on rare occasions the parity-check symbol will be the number represented by a 1 in bit 17. When this parity-check symbol occurs, change it to any 16-bit number, thereby making an error in the encoding. The Reed–Solomon code is intended to correct errors, and forcing an occasional error in the encoder will decrease by a trivial amount the power of the code to correct random errors in the channel.

4.5   Conjugacy Constraints and BCH Codes

Whenever the symbol field F does not contain an element of order n, then the Reed–Solomon code over F of blocklength n does not exist because there is no Fourier transform of blocklength n in F. However, one may still find a Fourier transform of blocklength n by entering the extension field F^m. This is how the Bose–Chaudhuri–Hocquenghem (BCH) codes arise. The code is in the field F, and the transform is in an extension field of F.

For example, it is well known that the real field does not contain an nth root of unity (except when $n = 2$). This is why Fourier transforms of real-valued signals are in the complex extension field. Real-valued Reed–Solomon codes do not exist but real-valued BCH codes do exist with spectra in the complex field.

In a Galois field $GF(q)$, if an nth root of unity does not exist then a Reed–Solomon code of blocklength n does not exist in $GF(q)$. If an nth root of unity exists in $GF(q^m)$, an extension field of $GF(q)$, then a BCH code of blocklength n does exist in $GF(q)$.

In summary, a Fourier transform of blocklength n over $GF(q)$ may lie in $GF(q)$; it may lie in the extension field $GF(q^m)$; or it may not exist. In the first case there are Reed–Solomon codes of blocklength n over $GF(q)$; in the second case there are BCH codes of blocklength n over $GF(q)$.

Let F be a field, and let ω be an element of order n in the extension field F^m. The BCH code of blocklength n with symbols in F is the set of all vectors $\mathbf{c}$ in F whose transform in F^m satisfies $C_k = 0$ for $k = k_0, \ldots, k_0 + 2t - 1$. This is the same as the definition of a Reed–Solomon code except the transform is in the extension field F^m.

The set of codewords of a BCH code over F is contained in the set of codewords of a Reed–Solomon code over F^m. This is because F is contained in F^m and a vector of blocklength n over F with $C_k = 0$ for $k = 1, \ldots, 2t$ is an element of the set of vectors of blocklength n over F^m with $C_k = 0$ for $k = 1, \ldots, 2t$. The same would be true if a value of k_0 other than $k_0 = 1$ were chosen so that $C_k = 0$ for $k = k_0, \ldots, k_0 + 2t - 1$.

To see how the relation between F and F^m dictates the properties of the BCH code, we shall first study BCH codes over the real field. Recall that the transform of a real-valued vector $\mathbf{c}$ solves the conjugacy constraint

$$C_k = C^*_{n-k}.$$

Suppose we set C_k equal to zero for $k = 1, \ldots, 2t$. The conjugacy constraint then requires that for $k = 1, \ldots, 2t$, C_{n-k} equals zero also. Therefore only $n - 4t$ components of the spectrum are actually allowed to be nonzero. It is

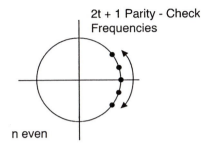
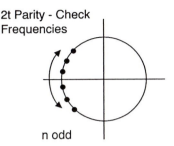

FIGURE 4.2. Complex zeros for real-valued BCH codes.

much better to choose $C_k = 0$ for $k = -t, \ldots, t$ (recalling that $C_{-1} = C_{n-1}$ because indices are modulo n). This gives the $2t+1$ parity-check frequencies for a t-error-correcting code as shown in Figure 4.2. This code has one extra parity-check frequency beyond those needed to correct t errors; it will correct t errors and detect $t + 1$ errors.

In the time domain, the encoding rule is

$$c(x) = g(x)d(x),$$

where $d(x)$ is a data polynomial of degree $n - 2t - 2$ with real coefficients and the generator polynomial is

$$g(x) = (x - \omega^{-t})(x - \omega^{-t+1}) \ldots (x - \omega^{t-1})(x - \omega^t).$$

It has real coefficients because it satisfies the conjugacy constraints. Therefore $n - 2t - 1$ real numbers are encoded into the n real numbers specified by the codeword. The original vector can be recovered even in the presence of t errors due to impulsive noise or burst noise in the channel.

If n is odd, we can choose the parity-check frequencies symmetrically about the index $\frac{n}{2}$. This way we satisfy the conjugacy requirement with only $2t$ parity-check frequencies with

$$k = \frac{n+1}{2} - t, \ldots, \frac{n-1}{2} + t.$$

Figure 4.2 illustrates this choice of parity-check frequencies.

In a Galois field, a BCH code again is like a Reed–Solomon code except that symbol field $GF(q)$ is a subfield of $GF(q^m)$, in which the Fourier transform is contained. For a binary code, the symbol field is $GF(2)$ and there are no Fourier transforms in $GF(2)$. To get a Fourier transform, one must go into an extension field $GF(2^m)$ for some m. For each m there are Fourier transforms for each blocklength n that divides $2^m - 1$. These are the possible blocklengths for binary-valued BCH codes.

To obtain a BCH code, set $C_k = 0$ for $k = 1, \ldots, 2t$ (or possibly for $k = k_0, \ldots, k_0 + 2t - 1$) and choose all such vectors $\mathbf{C}$ in the transform domain such that the inverse Fourier transform $\mathbf{c}$ has components only in $GF(2)$. To assure this latter condition, we can use the conjugacy constraint discussed in Section 3.4. This says that a vector $\mathbf{c}$ lies in the Galois field $GF(q)$ if and only if the transform $\mathbf{C}$ in $GF(q^m)$ satisfies

$$C_k^q = C_{((qk))}.$$

Hence, to obtain a BCH code over $GF(q)$, we need the two constraints on the spectrum

$$C_k = 0, \qquad\qquad k = 1, \ldots, 2t$$
$$C_k^q = C_{((qk))}, \qquad k = 0, \ldots, n - 1.$$

Except for the second constraint, everything about Reed–Solomon codes also applies to BCH codes even when the BCH codes are defined over the rational field.

For example, take $n = 7$ so the extension field is $GF(2^3)$, and take $t = 1$. This requires that $C_1 = C_2 = 0$. The conjugacy constraint requires that $C_4 = C_2^2 = 0$, that $C_0 = C_0^2$, and that $C_5 = C_6^2 = C_3^4$. Hence C_0 can only be 0 or 1 while C_5 and C_6 are determined by C_3. Consequently, one bit specifies C_0 and three bits specify C_3. The BCH (7,4) code so constructed is identical to the Hamming (7,4) code. Figure 4.3 shows the codewords of the Hamming (7,4) code as well as their transforms. This code can be encoded in the time domain as

$$c(x) = g(x)d(x),$$

using the generator polynomial

$$g(x) = (x - \omega)(x - \omega^2)(x - \omega^4).$$

The first two zeros of $g(x)$ ensure that $C_1 = C_2 = 0$, as is required to correct one error. The third zero of $g(x)$ is required in order to satisfy the conjugacy constraint. It produces a generator polynomial with binary coefficients.

Another example is the (15,7) double-error-correcting binary BCH code. Let ω have order 15 in the extension field $GF(16)$. The generator polynomial for this code is

$$g(x) = (x - \omega)(x - \omega^2)(x - \omega^3)(x - \omega^4)(x - \omega^8)(x - \omega^6)(x - \omega^{12})(x - \omega^9).$$

Frequency-Domain Codewords							Time-Domain Codewords						
C_0	C_1	C_2	C_3	C_4	C_5	C_6	c_0	c_1	c_2	c_3	c_4	c_5	c_6
0	0	0	0	0	0	0	0	0	0	0	0	0	0
0	0	0	ω^0	0	ω^0	ω^0	1	1	1	0	1	0	0
0	0	0	ω^1	0	ω^4	ω^2	0	0	1	1	1	0	1
0	0	0	ω^2	0	ω^1	ω^4	0	1	0	0	1	1	1
0	0	0	ω^3	0	ω^5	ω^6	1	1	0	1	0	0	1
0	0	0	ω^4	0	ω^2	ω^1	0	1	1	1	0	1	0
0	0	0	ω^5	0	ω^6	ω^3	1	0	0	1	1	1	0
0	0	0	ω^6	0	ω^3	ω^5	1	0	1	0	0	1	1
1	0	0	0	0	0	0	1	1	1	1	1	1	1
1	0	0	ω^0	0	ω^0	ω^0	0	0	0	1	0	1	1
1	0	0	ω^1	0	ω^4	ω^2	1	1	0	0	0	1	0
1	0	0	ω^2	0	ω^1	ω^4	1	0	1	1	0	0	0
1	0	0	ω^3	0	ω^5	ω^6	0	0	1	0	1	1	0
1	0	0	ω^4	0	ω^2	ω^1	1	0	0	0	1	0	1
1	0	0	ω^5	0	ω^6	ω^3	0	1	1	0	0	0	1
1	0	0	ω^6	0	ω^3	ω^5	0	1	0	1	1	0	0

FIGURE 4.3. Spectral description of the Hamming (7,4) code.

The first four zeros of $g(x)$ ensure that $C_1 = C_2 = C_3 = C_4 = 0$ as is required to correct two errors. The last four zeros of $g(x)$ are required in order to satisfy the conjugacy constraint. It produces a generator polynomial with binary coefficients.

Finally, there are also BCH codes in the field $\mathbf{Q}$ with transforms in $\mathbf{Q}^m$ (and even BCH codes in $\mathbf{Q}^m$ with transforms in $(\mathbf{Q}^m)^{m'}$). The conjugacy constraint that we need is that the inverse Fourier transform $\mathbf{v}$ is a vector over the base field $\mathbf{Q}$ if and only if the components of the transform satisfy the polynomial relationship.

$$V_{mk}(x) = V_k(x^m) \pmod{p(x)}.$$

The BCH codes in $\mathbf{Q}$ are obtained by forcing the transform to satisfy this condition.

5

Fast Algorithms for Convolution

The most important structure in digital signal processing is the nonrecursive digital filter known as the *finite impulse response* (FIR) filter. The FIR filter is simply a tapped delay line. It convolves the input datastream with the filter taps. An incoming datastream of digital samples enters the filter and an outgoing datastream of digital samples leaves the filter. The datastreams entering and leaving the filter are so long as to appear infinite. Perhaps thousands or even millions of samples pass through the filter per second, and this may continue indefinitely.

By using a fast algorithm one can greatly reduce the number of digital parts (or computer operations) that go into building a FIR filter. However the relationship between the parts will now be more intricate. If the nonrecurring cost of design time is unimportant compared to the hardware cost, then one should use a fast algorithm. If the design time is significant, then perhaps the obvious implementation of the tapped delay line should be used.

5.1 Convolution by Blocks

In the field F, a FIR filter with input starting at sample index 0 is described by the acyclic convolution

$$s_i = \sum_{k=0}^{L-1} g_k d_{i-k}, \qquad i = 0, 1, 2, \ldots,$$

where $d_0, d_1, d_2, \ldots$ is the input data sequence, $s_0, s_1, s_2, \ldots$ is the output data sequence, and $g_0, g_1, \ldots, g_{L-1}$ are the L tap weights of the FIR filter. The field F in which the FIR filter is defined is arbitrary, although problems of digital signal processing are usually in the real field or the complex field.

Fast algorithms for computing the output of a FIR filter break the input datastream into short sections of perhaps a few hundred samples. One section at a time is processed to compute a section of the output datastream. The output datastream is then assembled from the computed sections.

The input data are clocked into a buffer memory until an input section is filled. Then that section of the input datastream is processed by an efficient, highly optimized routine called a fast convolution algorithm. The fast algorithm works on the data of that input section as a batch, placing the result in an output buffer. The words in the output buffer are sequentially read out one by one onto the output line. The internal operations of the algorithm may look very different from the functional description of the FIR filter. From the input/output point of view, however, the implementation looks just as if it were a tapped delay line, except for a fixed delay.

For example, to pass two data samples d_0, d_1 through a two-tap FIR filter with constant tap weights g_0, g_1, the computation must produce three filter outputs given by

$$
\begin{aligned}
s_0 &= d_0 g_0, \\
s_1 &= d_1 g_0 + d_0 g_1, \\
s_2 &= d_1 g_1.
\end{aligned}
$$

This set of equations, in the obvious form, uses four multiplications and one addition.

A fast algorithm for the computation is

$$
\begin{aligned}
s_0 &= d_0 G_0, \\
s_1 &= (d_0 + d_1) G_2 - d_0 G_0 - d_1 G_1, \\
s_2 &= d_1 G_1,
\end{aligned}
$$

where the "precomputed" constants are $G_0 = g_0$, $G_1 = g_1$, and $G_2 = g_0 + g_1$. For each block of two data inputs, the fast algorithm uses three multiplications and three additions to compute a block of two data outputs. It is an improvement whenever the digital circuitry does two additions more easily than one multiplication.

This algorithm computes one section of three outputs from one section of two inputs. The next input section is processed by the same algorithm simply by incrementing the indices. Thus

$$
\begin{aligned}
s_2 &= d_2 G_0, \\
s_3 &= (d_2 + d_3) G_2 - d_2 G_0 - d_3 G_1, \\
s_4 &= d_3 G_1.
\end{aligned}
$$

This input section also makes a contribution to s_2 that must be added to the contribution from the first section. This kind of algorithm is called *overlap-add sectioning*, referring to the fact that the output sections overlap and the computed contribution in the overlapping segments must be added together to get the true output.

An alternative approach, called *overlap-save sectioning*, is to compute only the nonoverlapping portion of the output. Then a section is defined at the output as

$$
\begin{aligned}
s_0 &= d_0 g_0 + d_{-1} g_1, \\
s_1 &= d_1 g_0 + d_0 g_1,
\end{aligned}
$$

where the extraneous input data sample d_{-1} has been appended to lengthen the input section to three points, as is needed for subsequent input sections. A fast algorithm is

$$
\begin{aligned}
s_0 &= d_0(g_0 + g_1) + (d_{-1} - d_0)g_1, \\
s_1 &= d_0(g_0 + g_1) + (d_1 - d_0)g_0,
\end{aligned}
$$

which requires three multiplications and four additions. The next output section is

$$
\begin{aligned}
s_2 &= d_2(g_0 + g_1) + (d_1 - d_2)g_1, \\
s_3 &= d_2(g_0 + g_1) + (d_3 - d_2)g_0.
\end{aligned}
$$

Whereas the overlap-add sectioning uses nonoverlapping sections at the input of the computation, the overlap-save sectioning uses nonoverlapping sections at the output of the computation.

Of course, no one needs a fast algorithm for a filter with only two taps. It is discussed here only because it is small enough to serve as a simple example. Practical FIR filters may have a hundred or more taps and may have millions of input samples per second. Developing fast algorithms for these problems takes some work.

One way to construct fast algorithms for longer filters is to iterate the algorithms for smaller filters. First observe that the two-tap filter sections can be written

$$
\begin{aligned}
\begin{bmatrix} s_1 \\ s_2 \end{bmatrix} &=
\begin{bmatrix} d_1 & d_0 \\ d_2 & d_1 \end{bmatrix}
\begin{bmatrix} g_0 \\ g_1 \end{bmatrix} \\
&=
\begin{bmatrix} 0 & 1 & 1 \\ 1 & 1 & 0 \end{bmatrix}
\begin{bmatrix} d_2 - d_1 & 0 & 0 \\ 0 & d_1 & 0 \\ 0 & 0 & d_0 - d_1 \end{bmatrix}
\begin{bmatrix} 1 & 0 \\ 1 & 1 \\ 0 & 1 \end{bmatrix}
\begin{bmatrix} g_0 \\ g_1 \end{bmatrix}.
\end{aligned}
$$

Then observe that the four-tap filter section

$$
\begin{bmatrix} s_3 \\ s_4 \\ s_5 \\ s_6 \end{bmatrix} =
\begin{bmatrix}
d_3 & d_2 & d_1 & d_0 \\
d_4 & d_3 & d_2 & d_1 \\
d_5 & d_4 & d_3 & d_2 \\
d_6 & d_5 & d_4 & d_3
\end{bmatrix}
\begin{bmatrix} g_0 \\ g_1 \\ g_2 \\ g_3 \end{bmatrix}
$$

can be block-partitioned as

$$
\begin{bmatrix} s_3 \\ s_4 \\ \hline s_5 \\ s_6 \end{bmatrix} =
\left[
\begin{array}{cc|cc}
d_3 & d_2 & d_1 & d_0 \\
d_4 & d_3 & d_2 & d_1 \\
\hline
d_5 & d_4 & d_3 & d_2 \\
d_6 & d_5 & d_4 & d_3
\end{array}
\right]
\begin{bmatrix} g_0 \\ g_1 \\ g_2 \\ g_3 \end{bmatrix},
$$

which has the form

$$
\begin{bmatrix} S_1 \\ S_2 \end{bmatrix} = \begin{bmatrix} D_1 & D_0 \\ D_2 & D_1 \end{bmatrix} \begin{bmatrix} G_0 \\ G_1 \end{bmatrix}
$$

$$
= \begin{bmatrix} 0 & I & -I \\ I & I & 0 \end{bmatrix} \begin{bmatrix} D_2 - D_1 & 0 & 0 \\ 0 & D_1 & 0 \\ 0 & 0 & D_1 - D_0 \end{bmatrix}
$$

$$
\begin{bmatrix} I & 0 \\ I & I \\ 0 & I \end{bmatrix} \begin{bmatrix} G_0 \\ G_1 \end{bmatrix}.
$$

The last line follows by observing that the block-partitioned equation has the same form as the two-tap filter section. The four-tap filter section can be computed with three 2×2 matrix multiplications, each of which has the structure of a two-tap filter section and can be computed with three multiplications. Thus by iterating the two-tap filter section, a four-tap filter section can be computed with nine multiplications.

In general, a *radix-2* filter section consisting of 2^m outputs of a 2^m-tap filter can be iteratively computed using 3^m multiplications. In a similar way, an algorithm for a three-tap filter section can be designed with five multiplications. Therefore, a *radix-3* filter section consisting of $n = 3^m$ outputs of a 3^m-tap filter can be iteratively computed using 5^m multiplications.

Because $5^m = n^{\log_3 5}$ the number of multiplications eventually grows quickly compared to $n \log_2 n$. Thus for large n, the algorithms for iterated filter sections will not compete with methods that require on the order of $n \log_2 n$ multiplications. Such methods use the Fourier transform in conjunction with the convolution theorem.

5.2   Fast Algorithms for Cyclic Convolution

Algorithms for large filter sections can be constructed using algorithms for cyclic convolutions. Because a filter section has the form of an acyclic convolution, a cyclic convolution gives the wrong answer but is easy to compute. Therefore either the input data sequence or the tap sequence is padded with zeros and a cyclic convolution blocklength longer than the filter section is used. The output of the cyclic convolution is then adjusted by a side computation to put it in the form of a filter section.

Algorithms for cyclic convolution are of two kinds: algorithms in the time domain and algorithms in the frequency domain based on the Fourier transform and the convolution theorem. An algorithm in the frequency domain uses a fast algorithm for computing the Fourier transform. (Such algorithms are discussed in Chapter 7.) The two vectors to be cyclically convolved are transformed, and the spectrum is multiplied componentwise and inverse transformed. If $M(n)$ represents the number of multiplications in the field F needed to compute a Fourier transform, then the computation of the cyclic convolution takes $3M(n) + n$ multiplications in the field F because there are three Fourier transforms (two direct and one inverse) and n products of spectral components.

Fast algorithms for cyclic convolution also can be constructed in the time domain. Fast algorithms for cyclic convolution of small blocklength can be constructed by using the Chinese remainder theorem for polynomials and those algorithms can be combined by using the Chinese remainder theorem for integers to get a fast algorithm for cyclic convolution of large blocklength. Because a filter section is an acyclic convolution, a cyclic convolution is a poor match for a filter section but, surprisingly, is used a great deal to compute filter sections. Even though the acyclic convolution must be forced to make it look like a cyclic convolution, it can be an advantage because there are some very good fast algorithms for cyclic convolution.

By using the Chinese remainder theorem (discussed in Section 2.2) we can turn a cyclic convolution into a two-dimensional cyclic convolution. Suppose that $n = n'n''$, where n' and n'' are relatively prime. Replace the indices i and k by double indices (i', i'') and (k', k'') given by

$$\begin{aligned} i' &= i \quad (\bmod\ n'), \\ i'' &= i \quad (\bmod\ n'') \end{aligned}$$

and

$$\begin{aligned} k' &= k \quad (\bmod\ n'), \\ k'' &= k \quad (\bmod\ n''). \end{aligned}$$

The original indices can be recovered from the new indices by the inverse statement of the Chinese remainder theorem:

$$\begin{aligned} i &= N''n''i' + N'n'i'' \qquad (\bmod\ n), \\ k &= N''n''k' + N'n'k'' \qquad (\bmod\ n), \end{aligned}$$

where N' and N'' are those integers that satisfy

$$N'n' + N''n'' = 1.$$

The convolution

$$s_i = \sum_{k=0}^{n-1} g_{((i-k))} d_k$$

now can be rewritten as

$$s_{N''n''i'+N'n'i''} = \sum_{k'=0}^{n'-1} \sum_{k''=0}^{n''-1} g_{N''n''(i'-k')+N'n'(i''-k'')} d_{N''n''k'+N'n'k''}.$$

The double summation on k' and k'' is equivalent to the single summation on k because it picks up the same terms. The two-dimensional arrays $\mathbf{d}$,

g, and **s**, with indices $k' = 0, \ldots, n' - 1$ and $k'' = 0, \ldots, n'' - 1$, are now defined by

$$
\begin{aligned}
d_{k'k''} &= d_{N''n''k'+N'n'k''}, \\
g_{k'k''} &= g_{N''n''k'+N'n'k''}, \\
s_{k'k''} &= s_{N''n''k'+N'n'k''},
\end{aligned}
$$

so that the convolution becomes

$$
s_{i'i''} = \sum_{k'=0}^{n'-1} \sum_{k''=0}^{n''-1} g_{((i'-k'))((i''-k''))} d_{k'k''},
$$

where the first and second indices are interpreted modulo n' and modulo n'', respectively.

The $n'n''$-point cyclic convolution has been changed into a two-dimensional cyclic convolution. This will be more efficient only if efficient n'- and n''-point cyclic convolution algorithms are available. We will develop the Winograd cyclic convolution algorithms. To develop such algorithms, the Chinese remainder theorem for polynomials will be used. It is based on the structure of the ring of polynomials.

The ring of polynomials over a field F is quite similar in structure to the ring of integers. The division algorithm for polynomials says that given any polynomial $c(x)$ and nonzero polynomial $d(x)$ there exist unique polynomials $Q(x)$, called the *quotient polynomial*, and $s(x)$, called the *remainder polynomial*, such that $c(x) = d(x)Q(x) + s(x)$ and $\deg s(x) < \deg d(x)$.

The greatest common divisor of two polynomials $r(x)$ and $s(x)$, denoted $\mathrm{GCD}[r(x), s(x)]$, is the monic polynomial of largest degree that divides both. Two polynomials $r(x)$ and $s(x)$ are called *relatively prime* if $\mathrm{GCD}[r(x), s(x)] = 1$. A monic polynomial is a *prime polynomial* if it is relatively prime with every monic polynomial of smaller degree.

The *Euclidean algorithm* for polynomials is a recursive procedure to find the greatest common divisor of two polynomials $s(x)$ and $t(x)$. Let $s^{(0)}(x) = s(x)$ and $t^{(0)}(x) = t(x)$, where $\deg s(x) > \deg t(x)$. The recursion is

$$
\begin{bmatrix} s^{(r)}(x) \\ t^{(r)}(x) \end{bmatrix} = \begin{bmatrix} 0 & 1 \\ 1 & -Q^{(r)}(x) \end{bmatrix} \begin{bmatrix} s^{(r-1)}(x) \\ t^{(r-1)}(x) \end{bmatrix}
$$

as long as $t^{(r-1)}(x)$ is nonzero, and where

$$
Q^{(r)}(x) = \left\lfloor \frac{s^{(r-1)}(x)}{t^{(r-1)}(x)} \right\rfloor
$$

is the quotient obtained when $s^{(r-1)}(x)$ is divided by $t^{(r-1)}(x)$. When $t^{(r-1)}(x)$ is zero, $\mathrm{GCD}[s(x), t(x)] = s^{(r-1)}(x)$.

The 2×2 matrix of the recursion has determinant equal to -1. Therefore the inverse of the matrix exists as a matrix of polynomials as does the inverse of a product of such matrices, and so

$$\begin{bmatrix} s^{(r-1)}(x) \\ t^{(r-1)}(x) \end{bmatrix} = \mathbf{A}(x) \begin{bmatrix} s(x) \\ t(x) \end{bmatrix},$$

where $\mathbf{A}(x)$ is a 2×2 matrix of polynomials. Consequently,

$$\mathrm{GCD}[s(x), t(x)] = a(x)s(x) + b(x)t(x)$$

for some polynomials $a(x)$ and $b(x)$. In particular, if $s(x)$ and $t(x)$ are relatively prime, the equation

$$a(x)s(x) + b(x)t(x) = 1$$

is satisfied by some pair of polynomials $a(x)$ and $b(x)$.

The *Chinese remainder theorem* for polynomials says that for any set of pairwise relatively prime polynomials $\{m_0(x), m_1(x), \ldots, m_{K-1}(x)\}$ the set of congruences

$$c(x) \equiv c_k(x) \pmod{m_k(x)}, \qquad k = 0, \ldots, K - 1$$

has exactly one solution of degree smaller than the sum of the degrees of the $m_k(x)$. The set of congruences can be inverted by the equation

$$c(x) = \sum_{k=0}^{K-1} c_k(x) N_k(x) M_k(x) \pmod{M(x)},$$

where $M(x) = \prod_k m_k(x)$, $M_k(x) = M(x)/m_k(x)$, and $N_k(x)$ solves

$$N_k(x) M_k(x) + n_k(x) m_k(x) = 1.$$

Such an $N_k(x)$ must exist because $\mathrm{GCD}[M_k(x), m_k(x)] = 1$.

The application of the Chinese remainder theorem to the construction of the Winograd algorithms for cyclic convolution of small blocklength is as follows. Let

$$s(x) = g(x)d(x) \pmod{x^n - 1}.$$

Let $x^n - 1$ be written in terms of its prime factors:

$$x^n - 1 = p^{(0)}(x)p^{(1)}(x) \ldots p^{(K-1)}(x).$$

By taking residues, the original polynomial product becomes a set of polynomial products

$$s^{(k)}(x) = g^{(k)}(x)d^{(k)}(x) \pmod{p^{(k)}(x)}$$

for $k = 0, \ldots, K - 1$. Then $s(x)$ can be recovered using

$$s(x) = \sum_{k=0}^{K-1} a^{(k)}(x) s^{(k)}(x) \qquad (\mathrm{mod}\ x^n - 1).$$

The polynomials $a^{(k)}(x)$ for $k = 0, \ldots, K - 1$ are given by the Chinese remainder theorem for polynomials.

For example, suppose $n = 8$ and the field is the real field. Then $x^8 - 1$ can be factored as

$$x^8 - 1 = (x - 1)(x + 1)(x^2 + 1)(x^4 + 1).$$

The residue polynomials of $d(x)$ modulo these prime factors are

$$
\begin{aligned}
d^{(0)}(x) &= d(x) &&(\mathrm{mod}\ x - 1), \\
d^{(1)}(x) &= d(x) &&(\mathrm{mod}\ x + 1), \\
d^{(2)}(x) &= d(x) &&(\mathrm{mod}\ x^2 + 1), \\
d^{(3)}(x) &= d(x) &&(\mathrm{mod}\ x^4 + 1).
\end{aligned}
$$

The residue polynomials of $g(x)$ have exactly the same form. Between them the residues of $d(x)$ have eight coefficients as does $d(x)$. The computation of the residues has the nature of a transform when it is written in the following concise form.

$$
\begin{bmatrix}
d_0^{(0)} \\
d_0^{(1)} \\
d_0^{(2)} \\
d_1^{(2)} \\
d_0^{(3)} \\
d_1^{(3)} \\
d_2^{(3)} \\
d_3^{(3)}
\end{bmatrix}
=
\begin{bmatrix}
1 & 1 & 1 & 1 & 1 & 1 & 1 & 1 \\
1 & -1 & 1 & -1 & 1 & -1 & 1 & -1 \\
1 & 0 & -1 & 0 & 1 & 0 & -1 & 0 \\
0 & 1 & 0 & -1 & 0 & 1 & 0 & -1 \\
1 & 0 & 0 & 0 & -1 & 0 & 0 & 0 \\
0 & 1 & 0 & 0 & 0 & -1 & 0 & 0 \\
0 & 0 & 1 & 0 & 0 & 0 & -1 & 0 \\
0 & 0 & 0 & 1 & 0 & 0 & 0 & -1
\end{bmatrix}
\begin{bmatrix}
d_0 \\
d_1 \\
d_2 \\
d_3 \\
d_4 \\
d_5 \\
d_6 \\
d_7
\end{bmatrix}.
$$

In this form, it is quite clear that the coefficients of the residue polynomials can be computed without multiplications.

The Chinese remainder theorem gives the inverse of this "transform." In polynomial form, it is

$$
\begin{aligned}
d(x) = {} & \frac{1}{8}(x^7 + x^6 + x^5 + x^4 + x^3 + x^2 + x + 1)d^{(0)}(x) \\
& -\frac{1}{8}(x^7 - x^6 + x^5 - x^4 + x^3 - x^2 + x - 1)d^{(1)}(x) \\
& -\frac{1}{4}(x^6 - x^4 + x^2 - 1)d^{(2)}(x) \\
& -\frac{1}{2}(x^4 - 1)d^{(3)}(x)
\end{aligned}
$$

and in matrix form

$$
\begin{bmatrix} d_0 \\ d_1 \\ d_2 \\ d_3 \\ d_4 \\ d_5 \\ d_6 \\ d_7 \end{bmatrix} = \frac{1}{8}
\begin{bmatrix}
1 & 1 & 2 & 0 & 4 & 0 & 0 & 0 \\
1 & -1 & 0 & 2 & 0 & 4 & 0 & 0 \\
1 & 1 & -2 & 0 & 0 & 0 & 4 & 0 \\
1 & -1 & 0 & -2 & 0 & 0 & 0 & 4 \\
1 & 1 & 2 & 0 & -4 & 0 & 0 & 0 \\
1 & -1 & 0 & 2 & 0 & -4 & 0 & 0 \\
1 & 1 & -2 & 0 & 0 & 0 & -4 & 0 \\
1 & -1 & 0 & -2 & 0 & 0 & 0 & -4
\end{bmatrix}
\begin{bmatrix} d_0^{(0)} \\ d_0^{(1)} \\ d_0^{(2)} \\ d_1^{(2)} \\ d_0^{(3)} \\ d_1^{(3)} \\ d_2^{(3)} \\ d_3^{(3)} \end{bmatrix}.
$$

The small constants here often can be buried by redefining the coefficients of the FIR filter $g(x)$.

It remains to discuss the task of multiplying the residue polynomials. These products are

$$
\begin{aligned}
s^{(0)}(x) &= g^{(0)}(x)d^{(0)}(x) & (\bmod\ x - 1), \\
s^{(1)}(x) &= g^{(1)}(x)d^{(1)}(x) & (\bmod\ x + 1), \\
s^{(2)}(x) &= g^{(2)}(x)d^{(2)}(x) & (\bmod\ x^2 + 1), \\
s^{(3)}(x) &= g^{(3)}(x)d^{(3)}(x) & (\bmod\ x^4 + 1).
\end{aligned}
$$

The first two products are simply multiplications of real numbers because the polynomials have degree zero. The third product has the form of complex multiplication and can be computed with three real multiplications and three real additions using the factorization.

$$
(a + jb)(c + jd) = [(a - b)d + a(c - d)] + j[(a - b)d + b(c + d)].
$$

The last polynomial product has the form of multiplication in $\mathbf{Q}^4$. It can be computed with seven real multiplications and 41 real additions or with nine real multiplications and 15 real additions. In total, twelve real multiplications are needed to compute the eight-point cyclic convolution.

For another example, suppose that again $n = 8$ but now the field is the complex field. Then $x^8 - 1$ can be factored as

$$
x^8 - 1 = (x - 1)(x + 1)(x - j)(x + j)(x^2 - j)(x^2 + j).
$$

Now there are six residue polynomials, each of degree not larger than 2. The coefficients of the residue polynomials $d^{(k)}(x)$ can be expressed in the form of a matrix equation:

$$
\begin{bmatrix}
d_0^{(0)} \\
d_0^{(1)} \\
d_0^{(2)} \\
d_0^{(3)} \\
d_0^{(4)} \\
d_1^{(4)} \\
d_0^{(5)} \\
d_1^{(5)}
\end{bmatrix}
=
\begin{bmatrix}
1 & 1 & 1 & 1 & 1 & 1 & 1 & 1 \\
1 & -1 & 1 & -1 & 1 & -1 & 1 & -1 \\
1 & j & -1 & -j & 1 & j & -1 & -j \\
1 & -j & -1 & j & 1 & -j & -1 & j \\
1 & 0 & j & 0 & -1 & 0 & -j & 0 \\
0 & 1 & 0 & j & 0 & -1 & 0 & -j \\
1 & 0 & -j & 0 & -1 & 0 & j & 0 \\
0 & 1 & 0 & -j & 0 & -1 & 0 & j
\end{bmatrix}
\begin{bmatrix}
d_0 \\
d_1 \\
d_2 \\
d_3 \\
d_4 \\
d_5 \\
d_6 \\
d_7
\end{bmatrix}.
$$

The inverse of this matrix can be developed using the inverse equations of the Chinese remainder theorem.

Residue polynomial multiplications $g^{(k)}(x)d^{(k)}(x)$ for $k = 0, 1, 2, 3$ are simply multiplications of complex numbers and each can be implemented with three real multiplications; for $k = 4$ and 5 the multiplications are products of polynomials of first degree with complex coefficients and each can be implemented with nine real multiplications. A total of 30 real multiplications is required to compute an eight-point cyclic convolution in the complex field.

5.3   Convolution of Integer Sequences

A major computational task of digital signal processing is convolution; it is the essence of a FIR filter and of correlation. All practical problems have a limited wordlength of, say, b bits. By temporarily rescaling the data, we may suppose that the convolution is a convolution of integer sequences (or of sequences of Gaussian integers). This convolution will produce another sequence of integers. If the wordlength of the computation is chosen carefully, there will be no overflow in the input or the output data of the computation.

To convolve integer sequences cyclically, we can embed the integers into any surrogate field that is large enough to hold them without overflow and also large enough to hold the integers comprising the result of the computation without overflow. Then we use the convolution theorem in that field. To compute the cyclic convolution $\mathbf{g}*\mathbf{d}$, we compute the Fourier transforms of $\mathbf{g}$ and $\mathbf{d}$, multiply componentwise $G_k D_k$ in the transform domain, and then take the inverse Fourier transform. One's choice of surrogate field depends on the ease of the computations in that field and on one's personal biases. Good choices are the field of complex numbers and the Galois fields $GF(2^{16} + 1)$ or $GF(2^{17} - 1)$. It is hard to argue that the complex field is intrinsically more natural than the other two for convolution of integer sequences. Any one of them can be chosen for the computation.

The acyclic convolution

$$e_i = \sum_{\ell=0}^{n-1} g_{i-\ell} h_\ell, \qquad i = 0, \ldots, n-1$$

or the cyclic convolution

$$e_i = \sum_{\ell=0}^{n-1} g_{((i-\ell))} h_\ell, \qquad i = 0, \ldots, n-1$$

can be replaced by

$$e_i = \sum_{\ell=0}^{n-1} g_{i-\ell} h_\ell \pmod{p}, \qquad i = 0, \ldots, n-1$$

or

$$e_i = \sum_{\ell=0}^{n-1} g_{((i-\ell))} h_\ell \pmod{p}, \qquad i = 0, \ldots, n-1,$$

where p is a prime. As long as p is larger than any of the (input or output) integers in the computation, the modulo-p operation is superfluous. Now, however, the computation is in $GF(p)$ and we can exploit any computational tricks that we can find in $GF(p)$. This includes using the convolution theorem in $GF(p)$ and a fast Fourier transform in $GF(p)$.

If $\sqrt{-1}$ does not exist in the field $GF(p)$, then the extension field $GF(p^2)$ exists with the familiar multiplication rule

$$(a + jb)(c + jd) = (ac - bd) + j(bc + ad)$$

and $GF(p^2)$ has Fourier transforms of every blocklength n that divides $p^2 - 1$. The complex integers can be embedded into $GF(p^2)$ and cyclic convolutions of sequences of complex integers can be computed in the surrogate field $GF(p^2)$ using the convolution theorem and a fast Fourier transform algorithm in that field.

Even to convolve sequences of (real) integers, the field $GF(p^2)$ may be appropriate because the blocklengths of the Fourier transforms are attractive. For example, $GF(2^{17} - 1)$ has no radix-2 Fourier transforms but $GF((2^{17} - 1)^2)$ has all radix-2 blocklengths up to 2^{18}. In the real field, there is a standard trick to compute two Fourier transforms at once using the conjugacy constraint in the complex field that

$$V_{n-k}^* = V_k$$

if $\mathbf{v}$ is real valued. The Fourier transform of $\mathbf{v} = \mathbf{a} + j\mathbf{b}$ is $\mathbf{V} = \mathbf{A} + j\mathbf{B}$ with components

$$V_k = A_k + jB_k.$$

Because A_k and B_k are each complex, in general, V_k cannot be simply decomposed into A_k and B_k by taking the real part and the imaginary part. However the conjugacy constraint gives

$$V^*_{n-k} = A_k - jB_k,$$

so we have

$$
\begin{aligned}
2A_k &= V_k + V^*_{n-k}, \\
2jB_k &= V_k - V^*_{n-k}.
\end{aligned}
$$

In this way, two real vectors **a** and **b** can be represented temporarily by the complex vector **v**; the two Fourier transforms **A** and **B** can then be retrieved from the computation of a single Fourier transform **V**.

In the field $GF(p^2)$ with $p = -1 \pmod 4$ the analogous conjugacy constraint is

$$V^*_k = V_{((pk))},$$

as was discussed in Section 3.4. Therefore if $\mathbf{v} = \mathbf{a} + j\mathbf{b}$ is a vector in $GF(p^2)$ formed from two vectors **a** and **b** in $GF(p)$, we have

$$
\begin{aligned}
2A_k &= V_k + V^*_{((pk))}, \\
2jB_k &= V_k - V^*_{((pk))}
\end{aligned}
$$

for the components of the Fourier transforms **A** and **B** in terms of the components of **V**. In this way two Fourier transforms of "real" vectors are computed by one Fourier transform of a "complex" vector.

One can also use a Fermat field such as $GF(2^{16} + 1)$ to compute convolutions of complex integer sequences even though an extension of the Fermat field has the wrong structure for complex multiplication. Let the polynomials $g(x)$ and $d(x)$ having complex integer coefficients be written in terms of their real and imaginary parts as

$$g(x) = g_R(x) + jg_I(x),$$

and

$$d(x) = d_R(x) + jd_I(x).$$

We wish to compute the convolution represented by the polynomial product

$$s(x) = g(x)d(x).$$

One way is to compute the four convolutions $g_R(x)d_R(x)$, $g_I(x)d_R(x)$, $g_R(x)d_I(x)$, and $g_I(x)d_I(x)$ separately. A better procedure is to define the following polynomials with coefficients in $GF(2^{16} + 1)$:

$$
\begin{aligned}
a(x) &= \frac{1}{2}(g_R(x) - 2^8 g_I(x))(d_R(x) - 2^8 d_I(x)), \\
b(x) &= \frac{1}{2}(g_R(x) + 2^8 g_I(x))(d_R(x) + 2^8 d_I(x))
\end{aligned}
$$

(where 2^8 is actually $\sqrt{-1}$ in $GF(2^{16}+1)$). The output polynomial $s(x)$ is then given by

$$
\begin{aligned}
s_R(x) &= (a(x) + b(x)), \\
s_I(x) &= 2^8 (a(x) - b(x)).
\end{aligned}
$$

The polynomial products needed to compute $a(x)$ and $b(x)$ require only that two convolutions in $GF(2^{16}+1)$ be computed in place of one complex integer convolution. Hence, the number of multiplications is reduced by a factor of 2.

5.4   Convolutions Using Residue Number Systems

Using the Chinese remainder theorem, an integer c can be represented by its residues modulo a set of relatively prime moduli $\{m_j \mid j = 0, \ldots, J-1\}$ provided c lies in the range $0 \le c < \prod_j m_j$. The representation $\{c_j \mid j = 0, \ldots, J-1\}$ is called a residue number representation of c.

Addition and multiplication are preserved in a residue number representation. That is, if $c = a+b$ and $\{a_j \mid j = 0, \ldots, J-1\}$, $\{b_j \mid j = 0, \ldots, J-1\}$, $\{c_j \mid j = 0, \ldots, J-1\}$ are the residue number representations of a, b, and c then

$$
c_j = a_j + b_j \qquad (\text{mod } m_j).
$$

Similarly, if $c = ab$ then

$$
c_j = a_j b_j \qquad (\text{mod } m_j).
$$

The only condition is that a, b, and c lie in the range of validity of that residue number system: $0 \le a < \prod_j m_j$. (If negative numbers are desired, then with no other change in the equations, the range of the residue number system can be reinterpreted as $-M' \le a < \prod_j m_j - M'$ for any M'.)

The advantage of the residue number system is that the residues are small integers and several multiplications of small integers can be simpler than one multiplication of large integers. The disadvantages are that the sum of the wordlengths of all residues is greater than the original wordlength of the natural representation, that low-order bits of the natural representation cannot be rounded while in the residue number representation, and that overhead computations are needed to move the data from the natural representation to the residue number representation and back again. The third disadvantage can be negligible if there is a large number of computations to be performed, as in filtering.

The acyclic convolution

$$
s_i = \sum_{k=0}^{L-1} g_k d_{i-k}, \qquad i = 0, 1, 2, \ldots
$$

can be converted into a set of J acyclic convolutions of residues:

$$s_i^{(j)} = \sum_{k=0}^{L-1} g_k^{(j)} d_{i-k}^{(j)} \quad (\text{mod } m_j), \qquad i = 0, 1, 2, \ldots \quad j = 0, \ldots, J-1,$$

where

$$
\begin{aligned}
g_i^{(j)} &= g_i \quad (\text{mod } m_j), \\
d_i^{(j)} &= d_i \quad (\text{mod } m_j), \\
s_i^{(j)} &= s_i \quad (\text{mod } m_j),
\end{aligned}
$$

and the m_j are a set of relatively prime moduli. The $s_i^{(j)}$ are computed individually and then combined using the inverse equation of the Chinese remainder theorem to recover s_i.

Each residue convolution can be computed by any convenient method. Each can be broken into sections convenient for that computation; each section can be converted into a cyclic convolution; and, if m_j is a prime, an FFT in $GF(m_j)$ can be used to do the cyclic convolution.

To compute complex convolutions a residue number decomposition can be used on the real part and on the imaginary part. An alternative method is to approximate the complex numbers by elements of $\mathbf{Z}[e^{j2\pi/8}]$ or a restricted set of algebraic integers such as $\mathbf{Z}[e^{j2\pi/8}]_6$. Let

$$\tilde{d}_i = a_{0i} + a_{1i}\omega + a_{2i}\omega^2 + a_{3i}\omega^3$$

and

$$\tilde{g}_i = b_{0i} + b_{1i}\omega + b_{2i}\omega^2 + b_{3i}\omega^3$$

be the elements of $\mathbf{Z}[e^{j2\pi/8}]_6$ approximating d_i and g_i. Then

$$\tilde{s}_i = \sum_{k=0}^{L-1} \tilde{g}_k \tilde{d}_{i-k} \quad (\text{mod } \omega^4 + 1)$$

is an element of $\mathbf{Z}[e^{j2\pi/8}]$ approximating s_i. Each multiplication is a multiplication in the ring $\mathbf{Z}[e^{j2\pi/8}]$; it has the form of a multiplication (modulo $\omega^4 + 1$) of polynomials in ω of degree 3.

5.5   Convolution of Polynomial Sequences

Two-dimensional convolutions are important in problems of image processing and can require massive amounts of computation. A two-dimensional convolution can be viewed as a one-dimensional convolution of polynomials. Simply treat each row of the array as a polynomial; the elements in the row are the coefficients of the polynomial. Then the two-dimensional array becomes a one-dimensional array of polynomials and the two-dimensional convolution becomes a one-dimensional convolution of polynomials. Thus a good algorithm for computing convolutions of polynomials would be useful.

The convolution theorem holds in the field $\mathbf{Q}^m$. Hence a cyclic convolution of vectors of polynomials $\mathbf{g}(x)$ and $\mathbf{d}(x)$ in $\mathbf{Q}^m$ has the form

$$s_i(x) = \sum_{k=0}^{n-1} g_{((i-k))}(x)d_k(x) \qquad (\text{mod } p(x)),$$

where $p(x)$ is the irreducible polynomial used to form $\mathbf{Q}^m$. The vector of polynomials $\mathbf{s}(x)$ can be computed efficiently with the aid of the convolution theorem. First take the Fourier transforms of $\mathbf{g}(x)$ and $\mathbf{d}(x)$ in $\mathbf{Q}^m$, then multiply componentwise in the $\mathbf{Q}^m$ frequency domain, and finally take the inverse Fourier transform. Computation of the Fourier transform in $\mathbf{Q}^m$ uses no nontrivial real multiplications. The spectral components, however, are now represented by polynomials. The multiplications in the frequency domain are multiplications in $\mathbf{Q}^m$; that is, polynomial multiplications modulo the polynomial $p(x)$.

$$S_k(x) = G_k(x)D_k(x) \qquad (\text{mod } p(x)).$$

Multiplication in the frequency domain requires n polynomial products modulo the polynomial $p(x)$. The method will have value only if the computation of the spectral products is significantly simpler to compute than the original two-dimensional convolution.

We shall discuss only the case where m is a power of 2 and $p(x) = x^m + 1$. Then a suitable blocklength for a Fourier transform is $2m$ because $\omega = x$ has order $2m$. The simplest case has of the form

$$s_i(x) = \sum_{k=0}^{3} g_{((i-k))}(x)d_k(x) \qquad (\text{mod } x^2 + 1),$$

where all polynomials have degree less than 2. (In fact, this is just an unconventional way of writing a cyclic convolution of complex numbers of blocklength 4.) The Fourier transform is

$$D_k(x) = \sum_{i=0}^{3} x^{ik} d_i(x),$$

where $x^2 = -1$. Replacing x by j puts this in the familiar form of a four-point "multiply-free" Fourier transform over the complex field.

$$\begin{bmatrix} D_0 \\ D_1 \\ D_2 \\ D_3 \end{bmatrix} = \begin{bmatrix} 1 & 1 & 1 & 1 \\ 1 & j & -1 & -j \\ 1 & -1 & 1 & -1 \\ 1 & -j & -1 & j \end{bmatrix} \begin{bmatrix} d_0 \\ d_1 \\ d_2 \\ d_3 \end{bmatrix}.$$

Now the variables are thought of as complex numbers rather than as polynomials.

For a larger example, take $\mathbf{Q}^4$. The field elements are polynomials of degree at most 7. To emphasize the analogy with the complex field the polynomials will be written in the variable j instead of x. Then $j^4 = -1$ and $j^8 = 1$. Thus there is an eight-point "multiply-free" Fourier transform.

$$
\begin{bmatrix} D_0 \\ D_1 \\ D_2 \\ D_3 \\ D_4 \\ D_5 \\ D_6 \\ D_7 \end{bmatrix} = \begin{bmatrix} 1 & 1 & 1 & 1 & 1 & 1 & 1 & 1 \\ 1 & j & j^2 & j^3 & -1 & -j & -j^2 & -j^3 \\ 1 & j^2 & -1 & -j^2 & 1 & j^2 & -1 & -j^2 \\ 1 & j^3 & -j^2 & j & -1 & -j^3 & j^2 & -j \\ 1 & -1 & 1 & -1 & 1 & -1 & 1 & -1 \\ 1 & -j & j^2 & -j^3 & -1 & j & -j^2 & j^3 \\ 1 & -j^2 & -1 & j^2 & 1 & -j^2 & -1 & j^2 \\ 1 & -j^3 & -j^2 & -j & -1 & j^3 & j^2 & j \end{bmatrix} \begin{bmatrix} d_0 \\ d_1 \\ d_2 \\ d_3 \\ d_4 \\ d_5 \\ d_6 \\ d_7 \end{bmatrix}.
$$

The variables can be thought of as "hypercomplex" numbers with eight parts.

For a larger example, let $n = 64$, then the cyclotomic polynomial is $x^{32} + 1$. An element of $\mathbf{Q}^m$ is a polynomial of degree 31 described by a list of 32 rational numbers. The Fourier transform is

$$
V_k(x) = \sum_{i=0}^{63} x^{ik} v_i(x) \qquad (\mathrm{mod}\ x^{32} + 1),
$$

which can be computed by a Cooley–Tukey FFT (described in Chapter 7) with $\frac{1}{2} 64 \log_2 64$ butterfly stages. Each butterfly stage multiplies a polynomial of degree 31 by a power of x (mod $x^{32} + 1$) and adds two polynomials of degree 31. There are no multiplications of real numbers. A 64-point cyclic convolution is mapped into the componentwise product

$$
S_k(x) = G_k(x)D_k(x) \qquad (\mathrm{mod}\ x^{32} + 1).
$$

The Fourier transforms are easy to compute but the spectral products are computationally expensive. The real multiplications have been moved from the Fourier transform to the spectral products by using the field $\mathbf{Q}^m$ instead of $\mathbf{C}$. There would be no advantage here when convolving sequences of integers because the computation starts as an n-point cyclic convolution of integers and is transformed into a set of n products of polynomials. But what happens if the problem starts as an n-point cyclic convolution of polynomials? Then it is still transformed into a set of n products of polynomials. This time there will be a significant simplification because having n polynomial products is simpler than an n-point cyclic convolution of polynomials just as having n integer products is simpler than an n-point cyclic convolution of integers. The representation in $\mathbf{Q}^m$ is not very good for computing cyclic convolutions of integers but is quite good for computing cyclic convolutions of polynomials (mod $x^m + 1$). This can be useful because two-dimensional cyclic convolutions are important in image processing, and a two-dimensional cyclic convolution can be expressed as a one-dimensional cyclic convolution of polynomials.

To finish the treatment, we must show how to embed a cyclic convolution into $\mathbf{Q}^m$. A two-dimensional cyclic convolution of rational numbers is given by

$$s_{ii'} = \sum_{\ell=0}^{n-1} \sum_{\ell'=0}^{n'-1} g_{((i-\ell))((i'-\ell'))} d_{\ell\ell'},$$

where the double parentheses denote modulo n and modulo n', respectively. We shall treat only the case where n and n' are both powers of 2 and n' is not larger than n. The two-dimensional cyclic convolution can be expressed alternatively as a convolution of polynomials

$$s_i(x) = \sum_{\ell=0}^{n-1} g_{((i-\ell))}(x) d_\ell(x) \qquad (\text{mod } x^{n'} - 1),$$

where the polynomials are simply the columns of the two-dimensional arrays expressed as polynomials; each element of the column gives one coefficient of the polynomial. Specifically,

$$d_i(x) = \sum_{i'=0}^{n'-1} d_{ii'} x^{i'},$$

$$g_i(x) = \sum_{i'=0}^{n'-1} g_{ii'} x^{i'},$$

$$s_i(x) = \sum_{i'=0}^{n'-1} s_{ii'} x^{i'}.$$

We have seen how to compute a cyclic convolution of polynomials when the polynomial products are modulo the polynomial $x^m + 1$, but this cyclic convolution involves polynomial products modulo the polynomial $x^{n'} - 1$. To get into the right form, we break the problem down by taking residues. Hence define

$$d_i'(x) = d_i(x) \qquad (\text{mod } x^{n'/2} + 1),$$
$$d_i''(x) = d_i(x) \qquad (\text{mod } x^{n'/2} - 1).$$

These are defined as the remainders obtained when $d_i(x)$ is divided by $x^{n'/2} + 1$ and $x^{n'/2} - 1$, respectively. These remainders are quickly found by replacing $x^{r+n'/2}$ by $\pm x^r$ whenever x appears as a power at least as large as $n'/2$. This works because the remainder of $x^{r+n'/2}$ when divided by $x^{n'/2} \pm 1$ is just $\pm x^r$.

Similarly, define

$$g_i'(x) = g_i(x) \qquad (\bmod\ x^{n'/2} + 1),$$
$$g_i''(x) = g_i(x) \qquad (\bmod\ x^{n'/2} - 1)$$

and

$$s_i'(x) = s_i(x) \qquad (\bmod\ x^{n'/2} + 1),$$
$$s_i''(x) = s_i(x) \qquad (\bmod\ x^{n'/2} - 1).$$

Then the cyclic convolution can be decomposed into two subcomputations by dividing through the original equation by $x^{n'/2} \pm 1$ and keeping the remainder.

$$s_i'(x) = \sum_{\ell=0}^{n-1} g_{((i-\ell))}'(x) d_\ell'(x) \qquad (\bmod\ x^{n'/2} + 1),$$

$$s_i''(x) = \sum_{\ell=0}^{n-1} g_{((i-\ell))}''(x) d_\ell''(x) \qquad (\bmod\ x^{n'/2} - 1).$$

These pieces are put together using the Chinese remainder theorem to get $s_i(x)$:

$$s_i(x) = -\frac{1}{2}(x^{n'/2} - 1)s_i'(x) + \frac{1}{2}(x^{n'/2} + 1)s_i''(x).$$

The last equation involves no multiplications and is relatively trivial to compute.

We have decomposed an $n \times n'$ two-dimensional cyclic convolution into two equations. The equation for $s_i'(x)$ is a one-dimensional cyclic convolution in $\mathbf{Q}^{n'/2}$ of blocklength n. The equation for $s_i''(x)$ is an $n \times n'/2$ two-dimensional cyclic convolution. The first can be computed by using an FFT in $\mathbf{Q}^{n'/2}$ provided $n'/2$ is not larger than $n/2$. The second can be computed by using the same procedure again. It will be split in turn into two smaller pieces. The entire procedure can be formulated recursively. The terminating step of the recursion occurs as a convolution modulo the polynomial $(x-1)$, which is tantamount to taking the inner product of two scalar vectors.

For example, a 64×64 two-dimensional cyclic convolution requires three Fourier transforms in $\mathbf{Q}^{32}$ of blocklength 64 (which use no real multiplications) and 64 polynomial multiplications modulo the polynomial $x^{32} + 1$ plus a 64×32 two-dimensional cyclic convolution. In turn, the 32×64 cyclic convolution requires three Fourier transforms in $\mathbf{Q}^{32}$ of blocklength 32 and 32 polynomial multiplications modulo the polynomial $x^{32}+1$ plus a 32×32 two-dimensional cyclic convolution. In turn, the 32×32 two-dimensional cyclic convolution can be treated in the same way. This procedure can be repeated until the problem is exhausted.

6

Solving Toeplitz Systems

An $n \times n$ *Toeplitz matrix* is a square matrix in which element $a_{ij} = a_{i-j}$. An $n \times n$ *circulant matrix* is a square matrix in which element $a_{ij} = a_{((i-j))}$. A circulant matrix is a Toeplitz matrix. A *Toeplitz system of equations* is given by the matrix-vector equation $\mathbf{Af} = \mathbf{g}$. The computational task of solving the Toeplitz system of equations is the task of computing the vector $\mathbf{f}$ when given the vector $\mathbf{g}$ and the elements of the Toeplitz matrix on the left. One way to solve for $\mathbf{f}$ is to compute the matrix inverse. But to compute the matrix inverse for very large n may be impractical, both because of the amount of computation and because of problems of precision.

Toeplitz systems of equations occur in problems of spectral estimation, linear predictive coding of speech, and the decoding of error-control codes. The solution of a Toeplitz system of equations is a frequent task in seismic signal processing and is used extensively in petrochemical exploration. Toeplitz systems of equations with matrix dimensions of hundreds or even thousands may appear in applications. We may whimsically speculate that a historian of applied mathematics might one day inquire whether society would be different were there no efficient algorithm for the solution of Toeplitz systems of equations because the computations of petrochemical exploration would be orders of magnitude greater and possibly modern methods of petrochemical exploration could not exist.

6.1 The Sugiyama Algorithm

The Sugiyama algorithm inverts a Toeplitz system of equations of the form[1]

$$
\begin{bmatrix}
v_{n-1} & v_{n-2} & v_{n-3} & \cdots & v_0 \\
v_n & v_{n-1} & v_{n-2} & \cdots & v_1 \\
v_{n+1} & v_n & v_{n-1} & \cdots & v_2 \\
\vdots & & & & \\
v_{2n-2} & v_{2n-3} & v_{2n-4} & \cdots & v_{n-1}
\end{bmatrix}
\begin{bmatrix}
\Lambda_1 \\
\Lambda_2 \\
\Lambda_3 \\
\vdots \\
\Lambda_n
\end{bmatrix}
=
\begin{bmatrix}
-v_n \\
-v_{n+1} \\
-v_{n+2} \\
\vdots \\
-v_{2n-1}
\end{bmatrix}
$$

---

[1] In this section, it is more suggestive to index the components of the unknown vector from 1 to n, and to write the elements of $\mathbf{A}$ as v_0 to v_{2n-1}. Thus, the element v_{n-1} is on the diagonal of $\mathbf{A}$.

in any field F. The vector on the right is not arbitrary; it is made up of elements from the matrix on the left. The matrix is a Toeplitz matrix but no other structural requirements are imposed. It is not required to be a symmetric matrix. The matrix equation can be viewed as a description of a linear feedback shift register.

The Sugiyama algorithm solves the Toeplitz system of equations by employing the Euclidean algorithm for polynomials as a subalgorithm. Let

$$v(x) = \sum_{i=0}^{2n-1} v_i x^i,$$

$$\Lambda(x) = 1 + \sum_{i=1}^{n} \Lambda_i x^i$$

and consider the product $\Lambda(x)v(x)$. By inspection of the vector-matrix product above, we see that the ith coefficient of the product is equal to zero for $i = n, \ldots, 2n - 1$. Thus, as was asserted previously in Theorem 3.1.1, for some $p(x)$ and $g(x)$ with $\deg p(x) < n$,

$$\Lambda(x)v(x) = p(x) + x^{2n}g(x).$$

We will impose the additional requirement of a solution that $\mathrm{GCD}[p(x), \Lambda(x)] = 1$. This constraint is permissible because any factor common to both $p(x)$ and $\Lambda(x)$ must also be a factor of $g(x)$ and the common term can be divided out.

We now have replaced the problem of inverting the Toeplitz system of equations by the equivalent problem of finding $\Lambda(x)$ and $p(x)$ satisfying $\deg \Lambda(x) \le n$, $\deg p(x) \le n - 1$, and

$$\Lambda(x)v(x) = p(x) \qquad (\bmod\ x^{2n}).$$

Certainly, this equation has exactly one solution if the corresponding Toeplitz equation is invertible. If the corresponding Toeplitz equation is not invertible, then this equation need not have a solution.

The structure of the Euclidean algorithm for polynomials (discussed in Section 5.2) suggests a way to solve the equation for $\Lambda(x)$ and $p(x)$. From an inspection of the Euclidean algorithm,

$$Q^{(r)}(x) = \left\lfloor \frac{s^{(r-1)}(x)}{t^{(r-1)}(x)} \right\rfloor,$$

$$\mathbf{A}^{(r)}(x) = \begin{bmatrix} 0 & 1 \\ 1 & -Q^{(r)}(x) \end{bmatrix} \mathbf{A}^{(r-1)}(x),$$

$$\begin{bmatrix} s^{(r)}(x) \\ t^{(r)}(x) \end{bmatrix} = \begin{bmatrix} 0 & 1 \\ 1 & -Q^{(r)}(x) \end{bmatrix} \begin{bmatrix} s^{(r-1)}(x) \\ t^{(r-1)}(x) \end{bmatrix},$$

it is easy to see that there is an equation of polynomials of the form

$$\left[\begin{array}{c} s^{(r)}(x) \\ t^{(r)}(x) \end{array}\right] = \left[\begin{array}{cc} A_{11}^{(r)}(x) & A_{12}^{(r)}(x) \\ A_{21}^{(r)}(x) & A_{22}^{(r)}(x) \end{array}\right] \left[\begin{array}{c} s(x) \\ t(x) \end{array}\right]$$

so that

$$t^{(r)}(x) = A_{22}^{(r)}(x)t(x) \qquad (\mathrm{mod}\ s(x)),$$

which is the form of the equation being solved if we take $t(x) = v(x)$ and $s(x) = x^{2n}$. Such an equation holds for each r. To solve the problem we need to find an r for which $\deg t^{(r)}(x) \leq n-1$ and $\deg A_{22}^{(r)}(x) \leq n$, if such an r exists. Then these polynomials must equal the desired $\Lambda(x)$ and $p(x)$.

□ **Theorem 6.1.1** *Suppose the Euclidean algorithm for polynomials is exercised starting with $s^{(0)}(x) = x^{2n}$ and $t^{(0)}(x) = v(x)$. If $\bar{r}$ is the integer for which*

$$\begin{aligned} \deg t^{(\bar{r}-1)}(x) &\geq n, \\ \deg t^{(\bar{r})}(x) &< n, \end{aligned}$$

then

$$\begin{aligned} p(x) &= t^{(\bar{r})}(x)/A_{22}^{(\bar{r})}(0), \\ \Lambda(x) &= A_{22}^{(\bar{r})}(x)/A_{22}^{(\bar{r})}(0) \end{aligned}$$

satisfy the equation

$$p(x) = \Lambda(x)v(x) \qquad (\mathrm{mod}\ x^{2n})$$

with $\deg \Lambda(x) \leq n$, $\deg p(x) \leq n-1$, and $\Lambda_0 = 1$.

Proof The normalization ensures that $\Lambda_0 = 1$. The specified value of $\bar{r}$ exists and is unique because $\deg t^{(0)}(x) = 2n$, and the degree of $t^{(r)}(x)$ is strictly decreasing as r is increasing. By the definition of the specified $\bar{r}$, we have satisfied the first requirement,

$$\deg t^{(\bar{r})}(x) \leq n-1.$$

As r is increasing, the degree of $A_{22}^{(r)}(x)$ is increasing. We only need to show that

$$\deg A_{22}^{(\bar{r})}(x) \leq n.$$

This we prove by working with the inverse of the matrix $\mathbf{A}(x)$. First recall

$$\mathbf{A}^{(\bar{r})}(x) = \prod_{r=1}^{\bar{r}} \left[\begin{array}{cc} 0 & 1 \\ 1 & -Q^{(r)}(x) \end{array}\right],$$

where from this equation it is clear that $\deg A_{22}^{(\bar{r})}(x) > \deg A_{12}^{(\bar{r})}(x)$. Also recall that $\deg s^{(\bar{r})}(x) > \deg t^{(\bar{r})}(x)$. From these inequalities and the matrix equation

$$
\begin{bmatrix} s(x) \\ t(x) \end{bmatrix} = (-1)^r \begin{bmatrix} A_{22}^{(r)}(x) & -A_{12}^{(r)}(x) \\ -A_{21}^{(r)}(x) & A_{11}^{(r)}(x) \end{bmatrix} \begin{bmatrix} s^{(r)}(x) \\ t^{(r)}(x) \end{bmatrix},
$$

it is clear that $\deg s(x) = \deg A_{22}^{(\bar{r})}(x) + \deg s^{(\bar{r})}(x)$, and since $s^{(\bar{r})}(x) = t^{(\bar{r}-1)}(x)$, this becomes

$$
\begin{aligned}
\deg A_{22}^{(\bar{r})}(x) &= \deg s(x) - \deg t^{(\bar{r}-1)}(x) \\
&\leq 2n - n = n,
\end{aligned}
$$

where the inequality follows from the definition of $\bar{r}$. □

6.2   The Berlekamp–Massey Algorithm

The Berlekamp–Massey algorithm inverts the same Toeplitz system of equations as the Sugiyama algorithm. It is simpler computationally but is more difficult to derive.

The Toeplitz system of equations is interpreted as a description of a linear feedback shift register. Thus the task of inverting the Toeplitz system of equations is equivalent to the task of designing a linear feedback shift register that will generate the given sequence of filter outputs. As long as the Toeplitz matrix is invertible, there will be exactly one linear feedback shift register that satisfies the system of equations.

The Berlekamp–Massey algorithm computes the shortest linear feedback shift register $(\Lambda(x), L)$ that generates the given sequence. It computes the linear complexity L of the sequence and the feedback polynomial $\Lambda(x)$ that achieves the linear complexity. If the Toeplitz system has at least one solution, then the linear complexity is at most n. If the Toeplitz system is invertible, then $L = n$. If the Toeplitz system has no solutions, then $L > n$.

The design procedure is iterative. For each r, starting with $r = 1$, a minimum-length linear feedback shift register, denoted $(\Lambda^{(r)}(x), L_r)$, is designed for generating the sequence $v_0, \ldots, v_{r-1}$. At the start of iteration r, we will have constructed a list of shift registers

$$
\begin{aligned}
&(\Lambda^{(1)}(x), L_1), \\
&(\Lambda^{(2)}(x), L_2), \\
&\quad\vdots \\
&(\Lambda^{(r-1)}(x), L_{r-1}).
\end{aligned}
$$

The Berlekamp–Massey algorithm computes a new shortest-length linear feedback shift register $(\Lambda^{(r)}(x), L_r)$ that generates the sequence $v_0, \ldots,$

v_{r-2}, v_{r-1}, if possible by reusing the most recent shift register $(\Lambda^{(r-1)}(x), L_{r-1})$; otherwise remodeling that shift register by modifying the length and the tap weights.

At iteration r, the next output of the $(r-1)$th shift register is

$$\hat{v}_{r-1} = -\sum_{j=1}^{L_{r-1}} \Lambda_j^{(r-1)} v_{r-1-j},$$

which may or may not be equal to the desired output v_{r-1}. Let Δ_r be the difference between the desired output v_{r-1} and the actual output of the most recent shift register

$$
\begin{aligned}
\Delta_r &= v_{r-1} - \hat{v}_{r-1} \\
&= v_{r-1} + \sum_{j=1}^{L_{r-1}} \Lambda_j^{(r-1)} v_{r-1-j} \\
&= \sum_{j=0}^{L_{r-1}} \Lambda_j^{(r-1)} v_{r-1-j}.
\end{aligned}
$$

If Δ_r is zero, then set $(\Lambda^{(r)}(x), L_r) = (\Lambda^{(r-1)}(x), L_{r-1})$ and the rth iteration is complete. Otherwise, the feedback polynomial is modified as follows.

$$\Lambda^{(r)}(x) = \Lambda^{(r-1)}(x) + Ax^\ell \Lambda^{(m-1)}(x),$$

where A is a field element, ℓ is an integer, and $\Lambda^{(m-1)}(x)$ is one of the feedback polynomials appearing earlier on the list. Now with this new polynomial, let

$$
\begin{aligned}
\Delta_r' &= \sum_{j=0}^{L_r} \Lambda_j^{(r)} v_{r-1-j} \\
&= \sum_{j=0}^{L_r} \Lambda_j^{(r-1)} v_{r-1-j} + A \sum_{j=0}^{L_r} \Lambda_j^{(m-1)} v_{r-1-j-\ell},
\end{aligned}
$$

where $L_r = \max(L_{r-1}, \ell + L_{m-1})$.

We are ready to specify m, ℓ, and A to make $\Delta_r' = 0$. Choose an m smaller than r for which $\Delta_m \neq 0$, choose $\ell = r - m$, and choose $A = -\Delta_m^{-1}\Delta_r$. Then

$$\Delta_r' = \Delta_r - \frac{\Delta_r}{\Delta_m}\Delta_m = 0.$$

Consequently, the new shift register will indeed generate the sequence v_0, v_1, ..., v_{r-2}, v_{r-1}.

We still have not specified which of those m for which $\Delta_m \neq 0$ should be chosen, as we have not given a condition to ensure that each new shift register has minimum length. Theorem 6.2.4 asserts that if we choose m as the most recent iteration at which $L_m > L_{m-1}$, we will get a minimum-length shift register at every iteration, but this refinement will take some time to develop. Before stating and proving that theorem, we give several preliminary theorems.

□ **Theorem 6.2.1** *Let* $\mathbf{u}$ *and* $\mathbf{v}$ *be sequences of length* r. *If* $(\Lambda_1(x), L_1)$ *generates* $\mathbf{u}$ *and* $(\Lambda_2(x), L_2)$ *generates* $\mathbf{v}$, *then* $(\Lambda_1(x)\Lambda_2(x), L_1 + L_2)$ *generates* $\mathbf{u} + \mathbf{v}$.

Proof The proof is trivial if $L_1 + L_2 \geq r$ because then the shift-register contains the desired sequence $\mathbf{u} + \mathbf{v}$ as an initial condition. Suppose that $L_1 + L_2 < r$. Then

$$\Lambda_1(x)u(x) = p_1(x) + x^r g_1(x),$$
$$\Lambda_2(x)v(x) = p_2(x) + x^r g_2(x)$$

with $\deg p_1(x) < L_1$ and $\deg p_2(x) < L_2$. Consequently,

$$\Lambda_1(x)\Lambda_2(x)[u(x) + v(x)] = \Lambda_2(x)p_1(x) + \Lambda_1(x)p_2(x)$$
$$+ x^r[\Lambda_2(x)g_1(x) + \Lambda_1(x)g_2(x)].$$

The theorem is proved because $\deg[\Lambda_2(x)p_1(x) + \Lambda_1(x)p_2(x)] < L_1 + L_2$.
□

□ **Theorem 6.2.2 (Massey)** *Suppose that* $(\Lambda(x), L)$ *is the minimum-length shift register that generates* $(v_0, v_1, v_2, \ldots, v_{r-2})$ *but does not generate* $\mathbf{v} = (v_0, v_1, \ldots, v_{r-2}, v_{r-1})$. *Then* $L(\mathbf{v}) \geq \max[L, r - L]$, *where* $L(\mathbf{v})$ *is the linear complexity of* $\mathbf{v}$.

Proof (Maurer and Viscardi) The inequality to be proved is a combination of two inequalities:

$$L(\mathbf{v}) \geq L,$$
$$L(\mathbf{v}) \geq r - L.$$

The first inequality is obvious because the minimum linear feedback shift register that generates a sequence must also generate any beginning portion of that sequence.

Let $\mathbf{u} = (\mathbf{v}^{r-1}, \hat{v}_{r-1})$ denote the sequence of length r generated by $(\Lambda(x), L)$. By assumption, $\hat{v}_{r-1} \neq v_{r-1}$. Consequently,

$$\mathbf{v} - \mathbf{u} = (0, 0, \ldots, 0, \Delta),$$

where $\Delta = v_{r-1} - \hat{v}_{r-1}$. Using Theorem 6.2.1,

$$\begin{aligned} r &= L(\mathbf{v} - \mathbf{u}) \\ &\leq L(\mathbf{v}) + L(\mathbf{u}) \\ &= L(\mathbf{v}) + L, \end{aligned}$$

which completes the proof of the theorem.   □

If we can find a linear feedback shift register that satisfies the inequality of Theorem 6.2.2 with equality, then it must be of shortest length. The proof of Theorem 6.2.3 gives a construction for this shift register.

□ **Theorem 6.2.3** *Suppose that* $(\Lambda^{(i)}(x), L_i)$ *for* $i = 1, \ldots, r$ *is a sequence of minimum-length linear feedback shift registers such that* $\Lambda^{(i)}(x)$ *generates* $v_0, v_1, \ldots, v_{i-1}$. *If* $\Lambda^{(r)}(x) \neq \Lambda^{(r-1)}(x)$, *then*

$$L_r = \max[L_{r-1}, r - L_{r-1}].$$

Proof By Theorem 6.2.2, L_r cannot be smaller than the right side. We shall construct a shift register that generates the required sequence and whose length equals the right side; it must be a minimum-length shift register. The proof is by induction. We give a construction for a shift register satisfying the theorem, assuming that we have already constructed such shift registers for all $k \leq r - 1$. For $k = 1, \ldots, r - 1$, let $(\Lambda^{(k)}(x), L_k)$ be the minimum-length shift register that generates $v_0, \ldots, v_{k-1}$. Assume for the induction argument that

$$L_k = \max[L_{k-1}, k - L_{k-1}], \qquad k = 1, \ldots, r - 1,$$

whenever $\Lambda^{(k)}(x) \neq \Lambda^{(k-1)}(x)$. This is clearly true for $k = 1$ if v_0 is nonzero because $L_0 = 0$ and $L_1 = 1$. More generally, if v_{i-1} is the first nonzero in the given sequence, then $L_{i-1} = 0$ and $L_i = i$. The induction argument then begins at $k = i$.

At the end of iteration $r - 1$, let m denote the value k had at the most recent iteration step that required a length change. That is, m is that integer such that

$$L_{r-1} = L_m > L_{m-1}.$$

We now have

$$\sum_{i=0}^{L_{r-1}} \Lambda_i^{(r-1)} v_{j-i} = \begin{cases} 0, & j = L_{r-1}, \ldots, r - 2, \\ \Delta_r, & j = r - 1. \end{cases}$$

If $\Delta_r = 0$ then the shift register $(\Lambda^{(r-1)}(x), L_{r-1})$ also generates the first r symbols of the sequence so that

$$(\Lambda^{(r)}(x), L_r) = (\Lambda^{(r-1)}(x), L_{r-1}).$$

If $\Delta_r \neq 0$, then a new shift register must be designed. Recall that by the definition of m and by the induction hypothesis,

$$v_j + \sum_{i=1}^{L_{m-1}} \Lambda_i^{(m-1)} v_{j-i} = \begin{cases} 0, & j = L_{m-1}, \ldots, m - 2, \\ \Delta_m \neq 0, & j = m - 1 \end{cases}$$

and

$$
\begin{aligned}
L_{r-1} &= L_m = \max[L_{m-1}, m - L_{m-1}] \\
&= m - L_{m-1}
\end{aligned}
$$

because $L_m > L_{m-1}$. Now choose the new polynomial

$$
\Lambda^{(r)}(x) = \Lambda^{(r-1)}(x) - \Delta_r \Delta_m^{-1} x^{r-m} \Lambda^{(m-1)}(x).
$$

Because $\deg \Lambda^{(r-1)}(x) \le L_{r-1}$ and $\deg[x^{r-m}\Lambda^{(m-1)}(x)] \le r - m + L_{m-1}$,

$$
\begin{aligned}
\deg \Lambda^{(r)}(x) &\ge \max[L_{r-1}, r - m + L_{m-1}] \\
&\ge \max[L_{r-1}, r - L_{r-1}]
\end{aligned}
$$

By Theorem 6.2.2, if $\Lambda^{(r)}(x)$ generates $v_0, \ldots, v_{r-1}$ then $L_r = \max[L_{r-1}, r - L_{r-1}]$ which implies that the induction hypothesis

$$
L_r = \max[L_{r-1}, r - L_{r-1}]
$$

can be continued. It only remains to prove that the shift register $(\Lambda^{(r)}(x), L_r)$ generates the required sequence. This is done by direct computation of the difference between v_j and the shift register feedback.

$$
\begin{aligned}
&\sum_{i=0}^{L_{r-1}} \Lambda_i^{(r-1)} v_{j-i} - \Delta_r \Delta_m^{-1} \left[\sum_{i=0}^{L_{m-1}} \Lambda_i^{(m-1)} v_{j-r+m-i} \right] \\
&= \begin{cases} 0, & j = L_r, L_r + 1, \ldots, r - 2, \\ \Delta_r - \Delta_r \Delta_m^{-1} \Delta_m = 0, & j = r - 1. \end{cases}
\end{aligned}
$$

Hence the shift register $(\Lambda^{(r)}(x), L_r)$ generates $v_0, \ldots, v_{r-1}$. In particular, $(\Lambda^{(2n)}(x), L_{2n})$ generates $v_0, \ldots, v_{2n-1}$ and the theorem is proven. □

□ **Theorem 6.2.4 (Berlekamp–Massey Algorithm)** *In any field, let $v_0, \ldots, v_{2n-1}$ be given. Under the initial conditions $\Lambda^{(0)}(x) = 1$, $B^{(0)}(x) = 1$, $L_0 = 0$; let the following set of recursive equations be used to compute $\Lambda^{(r)}(x)$;*

$$
\Delta_r = \sum_{j=0}^{n-1} \Lambda_j^{(r-1)} v_{r-1-j},
$$

$$
L_r = \delta_r(r - L_{r-1}) + (1 - \delta_r)L_{r-1},
$$

$$
\begin{bmatrix} \Lambda^{(r)}(x) \\ B^{(r)}(x) \end{bmatrix} = \begin{bmatrix} 1 & -\Delta_r x \\ \delta_r \Delta_r^{-1} & (1 - \delta_r)x \end{bmatrix} \begin{bmatrix} \Lambda^{(r-1)}(x) \\ B^{(r-1)}(x) \end{bmatrix},
$$

$r = 1, \ldots, 2n$, where $\delta_r = 1$ if both $\Delta_r \ne 0$ and $2L_{r-1} \le r - 1$, and otherwise $\delta_r = 0$. Then $\Lambda^{(2n)}(x)$ is the locator polynomial for **v**.

Proof The proof of Theorem 6.2.4 follows from that of Theorem 6.2.3.
□

The number of multiplications in the rth iteration of the Berlekamp–Massey algorithm is approximately equal to twice the degree of $\Lambda^{(r)}(x)$. The degree of $\Lambda^{(r)}(x)$ is about $\frac{r}{2}$ and there are $2n$ iterations, so there are about $2n^2 = \sum_{r=0}^{2n} r$ multiplications and about the same number of additions. In brief, we say there are on the order of n^2 multiplications, or formally $O(n^2)$ multiplications in the Berlekamp–Massey algorithm.

6.3   Relationships Between Algorithms

The Berlekamp–Massey algorithm and the Sugiyama algorithm solve the same system of equations, so it is reasonable to inquire whether the algorithms have an internal relationship. Clearly, they are not the same, because the Berlekamp–Massey algorithm begins with polynomials of degree zero, and at each iteration increases the degree of the polynomial iterates while the Sugiyama algorithm begins with polynomials of large degree and at each iteration decreases the degree of the polynomial iterates. Nonetheless it is possible to recast the description of the Sugiyama algorithm so as to expose similarities in the structure.

The Berlekamp–Massey algorithm at each iteration contains the computation

$$\begin{bmatrix} \Lambda(x) \\ B(x) \end{bmatrix} \longleftarrow \begin{bmatrix} 1 & -\Delta x \\ \delta\Delta^{-1} & (1-\delta)x \end{bmatrix} \begin{bmatrix} \Lambda(x) \\ B(x) \end{bmatrix}$$

and the matrix takes one of the forms

$$\begin{bmatrix} 1 & 0 \\ 0 & x \end{bmatrix}, \quad \begin{bmatrix} 1 & -\Delta x \\ 0 & x \end{bmatrix}, \quad \text{or} \quad \begin{bmatrix} 1 & -\Delta x \\ \Delta^{-1} & 0 \end{bmatrix}.$$

The Sugiyama algorithm, on the other hand, contains the computation

$$\begin{bmatrix} s(x) \\ t(x) \end{bmatrix} \longleftarrow \begin{bmatrix} 0 & 1 \\ 1 & -Q^{(r)}(x) \end{bmatrix} \begin{bmatrix} s(x) \\ t(x) \end{bmatrix}.$$

Because $Q^{(r)}(x)$ need not have degree 1, this seems different than the structure of the Berlekamp–Massey algorithm. Furthermore, the Sugiyama algorithm has a variable number of iterations while the Berlekamp–Massey algorithm has a fixed number of iterations.

To restructure the Sugiyama algorithm let $N = \deg Q^{(r)}(x)$ and write

$$\begin{bmatrix} 0 & 1 \\ 1 & -Q^{(r)}(x) \end{bmatrix} = \begin{bmatrix} 0 & 1 \\ 1 & -Q_0 \end{bmatrix} \begin{bmatrix} 1 & -Q_1 x \\ 0 & 1 \end{bmatrix} \begin{bmatrix} 1 & -Q_2 x^2 \\ 0 & 1 \end{bmatrix}$$
$$\cdots \begin{bmatrix} 1 & -Q_N x^N \\ 0 & 1 \end{bmatrix}.$$

If the iteration counter counts each of these submatrices as an iteration then the Sugiyama algorithm has a fixed number of iterations and the 2×2 matrices more closely resemble those of the Berlekamp–Massey algorithm. It is possible to make the similarity even stronger by redefining the polynomials to make each matrix contain only the first power of x.

6.4   The Levinson and Durbin Algorithms

The *Levinson algorithm* is an efficient algorithm that can be used in any field F for solving the system of equations $\mathbf{Af} = \mathbf{g}$ when $\mathbf{A}$ is both symmetric and Toeplitz. The system of equations then has the form

$$
\begin{bmatrix}
a_0 & a_1 & \cdots & a_{n-2} & a_{n-1} \\
a_1 & a_0 & \cdots & a_{n-3} & a_{n-2} \\
\vdots & & & \vdots & \vdots \\
a_{n-2} & a_{n-3} & \cdots & a_0 & a_1 \\
a_{n-1} & a_{n-2} & & a_1 & a_0
\end{bmatrix}
\begin{bmatrix}
f_0 \\
f_1 \\
\vdots \\
f_{n-2} \\
f_{n-1}
\end{bmatrix}
=
\begin{bmatrix}
g_0 \\
g_1 \\
\vdots \\
g_{n-2} \\
g_{n-1}
\end{bmatrix}.
$$

The matrix and vectors have been partitioned to show an $(n-1) \times (n-1)$ Toeplitz system of equations with the same form. This is the repetitive structure upon which the Levinson algorithm is based. Each iteration will begin with a solution to the smaller problem and will then adjust the solution to append the new row at the bottom and the new column at the right.

An *exchange matrix* $\mathbf{J}$ is an $n \times n$ matrix with the form of an identity matrix but with rows written in reverse. It satisfies $\mathbf{J}^2 = \mathbf{I}$. Then

$$\mathbf{Af} = \mathbf{g}$$

can be rewritten as

$$\mathbf{JAJJf} = \mathbf{Jg}.$$

Then, because $\mathbf{A}$ is symmetric and Toeplitz, $\mathbf{JAJ} = \mathbf{A}$, so

$$\mathbf{AJf} = \mathbf{Jg}.$$

The Levinson algorithm is an iterative algorithm; we index the iterations by r. At step r, the Levinson algorithm begins with a solution $\mathbf{f}^{(r)}$ to the rth truncated problem

$$
\begin{bmatrix}
a_0 & a_1 & a_2 & \cdots & a_{r-1} \\
a_1 & a_0 & a_1 & \cdots & a_{r-2} \\
\vdots & & & & \\
a_{r-1} & a_{r-2} & a_{r-3} & \cdots & a_0
\end{bmatrix}
\begin{bmatrix}
f_0^{(r)} \\
f_1^{(r)} \\
\vdots \\
f_{r-1}^{(r)}
\end{bmatrix}
=
\begin{bmatrix}
g_0 \\
g_1 \\
\vdots \\
g_{r-1}
\end{bmatrix}.
$$

Clearly, even though $\mathbf{f}^{(r)}$ solves the truncated problem it is not a truncated version of the $\mathbf{f}$ that solves the original problem. The Levinson algorithm will iteratively update $\mathbf{f}^{(r)}$ to $\mathbf{f}^{(r+1)}$. In this way $\mathbf{f}^{(n)}$ is the required solution to the original problem.

In addition to $\mathbf{f}^{(r)}$, several working variables are iterated. These are the scalars α_r, β_r, and γ_r, and a vector $\mathbf{t}^{(r)}$ of length r. At the rth iteration the working variables α_r, β_r, and γ_r, and $\mathbf{t}^{(r)}$ are chosen so that the following side equation is satisfied.

$$
\begin{bmatrix}
a_0 & a_1 & \cdots & a_{r-1} \\
a_1 & a_0 & \cdots & a_{r-2} \\
\vdots & & & \\
a_{r-1} & a_{r-2} & \cdots & a_0
\end{bmatrix}
\begin{bmatrix}
t_0^{(r)} \\
t_1^{(r)} \\
\vdots \\
t_{r-1}^{(r)}
\end{bmatrix}
=
\begin{bmatrix}
\alpha_r \\
0 \\
\vdots \\
0
\end{bmatrix}.
$$

The iteration is constructed so as to perpetuate equations of the same form when the Toeplitz matrix is enlarged by appending a row and a column.

The iterative equations are derived by forcing two new equations to be satisfied. The first equation is developed by writing

$$
\left[
\begin{array}{cccc:c}
a_0 & a_1 & \cdots & a_{r-1} & a_r \\
a_1 & a_0 & \cdots & a_{r-2} & a_{r-1} \\
\vdots & & & & \\
a_{r-1} & a_{r-2} & & a_0 & a_1 \\
\hdashline
a_r & a_{r-1} & \cdots & a_1 & a_0
\end{array}
\right]
\begin{bmatrix}
t_0^{(r)} \\
t_1^{(r)} \\
\vdots \\
t_{r-1}^{(r)} \\
0
\end{bmatrix}
=
\begin{bmatrix}
\alpha_r \\
0 \\
\vdots \\
0 \\
\beta_r
\end{bmatrix},
$$

which defines γ_r. If $\beta_r = 0$ then $\mathbf{t}^{(r+1)}$ is the same as $\mathbf{t}^{(r)}$ but lengthened by appending a zero. If $\beta_r \neq 0$ then the exchange matrix $\mathbf{J}$ can be used to write

$$
\begin{bmatrix}
a_0 & a_1 & \cdots & a_{r-1} & a_r \\
a_1 & a_0 & \cdots & a_r & a_{r-1} \\
\vdots & & & & \\
a_{r-1} & a_{r-2} & \cdots & a_0 & a_1 \\
a_r & a_{r-1} & \cdots & a_1 & a_0
\end{bmatrix}
\begin{bmatrix}
0 \\
t_{r-1}^{(r)} \\
\vdots \\
t_1^{(r)} \\
t_0^{(r)}
\end{bmatrix}
=
\begin{bmatrix}
\beta_r \\
0 \\
\vdots \\
0 \\
\alpha_r
\end{bmatrix}.
$$

To derive an equation for updating $\mathbf{t}^{(r)}$ to $\mathbf{t}^{(r+1)}$ write

$$
\begin{bmatrix}
t_0^{(r+1)} \\
t_1^{(r+1)} \\
\vdots \\
t_{r-1}^{(r+1)} \\
t_r^{(r+1)}
\end{bmatrix}
= k_1
\begin{bmatrix}
t_0^{(r)} \\
t_1^{(r)} \\
\vdots \\
t_{r-1}^{(r)} \\
0
\end{bmatrix}
+ k_2
\begin{bmatrix}
0 \\
t_{r-1}^{(r)} \\
\vdots \\
t_1^{(r)} \\
t_0^{(r)}
\end{bmatrix},
$$

where k_1 and k_2 are constants yet to be chosen. To choose the constants observe that

$$
\begin{bmatrix}
a_0 & a_1 & \cdots & a_r \\
a_1 & a_0 & \cdots & a_{r-1} \\
\vdots & & & \\
a_{r-1} & a_{r-2} & \cdots & a_1 \\
a_r & a_{r-1} & \cdots & a_0
\end{bmatrix}
\begin{bmatrix}
t_0^{(r+1)} \\
t_1^{(r+1)} \\
\vdots \\
t_{r-1}^{(r+1)} \\
t_r^{(r+1)}
\end{bmatrix}
= k_1
\begin{bmatrix}
\alpha_r \\
0 \\
\vdots \\
0 \\
\beta_r
\end{bmatrix}
+ k_2
\begin{bmatrix}
\beta_r \\
0 \\
\vdots \\
0 \\
\alpha_r
\end{bmatrix}
$$

and we wish to find $\mathbf{t}^{(r+1)}$ so that the right side of this equation is zero in the last component. Therefore the constants k_1 and k_2 can be chosen in any way such that

$$0 = k_1 \beta_r + k_2 \alpha_r.$$

We arbitrarily choose $k_1 = \alpha_r$. (However, a different choice for k_1 may give different numerical accuracy.) Then

$$k_2 = -\beta_r$$

and

$$
\begin{aligned}
\alpha_{r+1} &= k_1 \alpha_r + k_2 \beta_r \\
&= \alpha_r^2 - \beta_r^2.
\end{aligned}
$$

We also need an equation for updating $\mathbf{f}^{(r)}$ to get $\mathbf{f}^{(r+1)}$. We start with the equation

$$
\left[
\begin{array}{cccc|c}
a_0 & a_1 & \cdots & a_{r-1} & a_r \\
a_1 & a_0 & \cdots & a_{r-2} & a_{r-1} \\
\vdots & & & & \\
a_{r-1} & a_{r-2} & \cdots & a_0 & a_1 \\
\hline
a_r & a_{r-1} & & a_1 & a_0
\end{array}
\right]
\begin{bmatrix}
f_0^{(r)} \\
f_1^{(r)} \\
\vdots \\
f_{r-1}^{(r)} \\
0
\end{bmatrix}
=
\begin{bmatrix}
g_0 \\
g_1 \\
\vdots \\
g_{r-1} \\
\gamma_r
\end{bmatrix},
$$

which defines γ_r. If $\gamma_r = g_r$ then $\mathbf{f}^{(r+1)}$ is the same as $\mathbf{f}^{(r)}$ but lengthened by appending a zero. Otherwise use the exchange matrix $\mathbf{J}$ to write

$$
\begin{bmatrix}
a_0 & a_1 & \cdots & a_{r-1} & a_r \\
a_1 & a_0 & \cdots & a_{r-2} & a_{r-1} \\
\vdots & & & & \\
a_{r-1} & a_{r-2} & & a_0 & a_1 \\
a_r & a_{r-1} & \cdots & a_1 & a_0
\end{bmatrix}
\begin{bmatrix}
0 \\
f_{r-1}^{(r)} \\
\vdots \\
f_0^{(r)}
\end{bmatrix}
=
\begin{bmatrix}
\gamma_r \\
g_{r-1} \\
\vdots \\
g_1 \\
g_0
\end{bmatrix}.
$$

To derive the rule for updating $\mathbf{f}^{(r)}$ to get $\mathbf{f}^{(r+1)}$ write

$$
\begin{bmatrix}
f_0^{(r+1)} \\
f_1^{(r+1)} \\
\vdots \\
f_{r-1}^{(r+1)} \\
f_r^{(r+1)}
\end{bmatrix}
=
\begin{bmatrix}
f_0^{(r)} \\
f_1^{(r)} \\
\vdots \\
f_{r-1}^{(r)} \\
0
\end{bmatrix}
+ k_3
\begin{bmatrix}
t_r^{(r+1)} \\
t_{r-1}^{(r+1)} \\
\vdots \\
t_1^{(r+1)} \\
t_0^{(r+1)}
\end{bmatrix}
$$

where k_3 is a constant yet to be chosen. Then

$$
\begin{bmatrix}
a_0 & a_1 & \cdots & a_r \\
a_1 & a_0 & \cdots & a_{r-1} \\
\vdots & & & \vdots \\
a_{r-1} & & & a_1 \\
a_r & & & a_0
\end{bmatrix}
\begin{bmatrix}
f_0^{(r+1)} \\
f_1^{(r+1)} \\
\vdots \\
f_{r-1}^{(r+1)} \\
f_r^{(r+1)}
\end{bmatrix}
=
\begin{bmatrix}
g_0 \\
g_1 \\
\vdots \\
g_r \\
\gamma_r
\end{bmatrix}
+ k_3
\begin{bmatrix}
0 \\
0 \\
\vdots \\
0 \\
\alpha_{r+1}
\end{bmatrix}.
$$

We want $\mathbf{f}^{(r+1)}$ such that the terms on the right sum to the vector $\mathbf{g}$. Therefore k_3 is to be chosen such that

$$
\gamma_r + k_3 \alpha_{r+1} = g_{r+1}.
$$

Thus,

$$
k_3 = \frac{g_{r+1} - \gamma_r}{\gamma_r^2 - \beta_r^2}.
$$

This completes the iteration.

The vectors $\mathbf{t}^{(r)}$ and $\mathbf{f}^{(r)}$ can be represented by polynomials

$$
\begin{aligned}
t(x) &= t_r^{(r)} x^r + t_{r-1}^{(r)} x^{r-1} + \ldots + t_1^{(r)} x + t_0^{(r)}, \\
f(x) &= f_r^{(r)} x^r + f_{r-1}^{(r)} x^{r-1} + \ldots + f_1^{(r)} x + f_0^{(r)},
\end{aligned}
$$

with the superscript r suppressed. There is no need actually to form the matrix $\mathbf{A}$ in the algorithm; it is enough to compute $f(x)$ and $t(x)$ as iterates by

$$
\begin{aligned}
t(x) &\leftarrow \alpha_r t(x) - \beta_r x^r t(x^{-1}), \\
f(x) &\leftarrow f(x) + \frac{g_r - \gamma_r}{\alpha_r^2 - \beta_r^2} x^r t(x^{-1}).
\end{aligned}
$$

The rth iteration has a complexity proportional to r and there are n passes. Therefore the complexity of the Levinson algorithm is proportional to n^2.

The Levinson algorithm holds in any field. In particular it holds just as stated in the complex field. However, a symmetric Toeplitz matrix in the complex field does not often arise in applications; a Hermitian Toeplitz matrix is much more common. The Levinson algorithm also holds in this case provided that one takes complex conjugates at the right points in the computation. It is easy to rework the derivation for this case.

The Levinson algorithm requires that the Toeplitz matrix is symmetric. A more efficient algorithm known as the *Durbin algorithm* can be used whenever the system of equations has the additional property that the vector on the right is made up of elements from the Toeplitz matrix in such a way that the system of equations takes the following form

$$
\left[
\begin{array}{cccc|c}
a_0 & a_1 & \cdots & a_{n-2} & a_{n-1} \\
a_1 & a_0 & & a_{n-3} & a_{n-2} \\
\vdots & & & & \vdots \\
a_{n-2} & & & a_0 & a_1 \\
\hline
a_{n-1} & & & a_1 & a_0
\end{array}
\right]
\left[
\begin{array}{c}
f_0 \\
f_1 \\
\vdots \\
f_{n-2} \\
\hline
f_{n-1}
\end{array}
\right]
= -
\left[
\begin{array}{c}
a_1 \\
a_2 \\
\vdots \\
a_{n-1} \\
\hline
a_n
\end{array}
\right].
$$

The matrix and vectors have again been blocked to show the repetitive bordering structure upon which the algorithm is based. The special property that the vector in the right column is made up of elements of the Toeplitz matrix allows the equations to be solved with half the work of the Levinson algorithm because only one polynomial needs to be iterated. The Durbin algorithm is useful because systems of Toeplitz equations of this form occur frequently in problems of spectral analysis.

At step r, the Durbin algorithm begins with a solution to the truncated problem

$$
\left[
\begin{array}{cccc}
a_0 & a_1 & \cdots & a_{r-1} \\
a_1 & a_0 & \cdots & a_{r-2} \\
\vdots & & & \\
a_{r-1} & a_{r-2} & \cdots & a_0
\end{array}
\right]
\left[
\begin{array}{c}
f_0^{(r)} \\
f_1^{(r)} \\
\vdots \\
f_{r-1}^{(r)}
\end{array}
\right]
= -
\left[
\begin{array}{c}
a_1 \\
a_2 \\
\vdots \\
a_r
\end{array}
\right].
$$

The next iteration begins with

$$
\left[
\begin{array}{ccccc|c}
a_0 & a_1 & a_2 & \cdots & a_{r-1} & a_r \\
a_1 & a_0 & a_1 & \cdots & a_{r-2} & a_{r-1} \\
\vdots & & & & & \\
a_{r-1} & a_{r-2} & & & a_0 & a_1 \\
\hline
a_r & a_{r-1} & & & a_1 & a_0
\end{array}
\right]
\left[
\begin{array}{c}
f_0^{(r)} \\
f_1^{(r)} \\
\vdots \\
f_{r-1}^{(r)} \\
\hline
0
\end{array}
\right]
= -
\left[
\begin{array}{c}
a_1 \\
a_2 \\
a_3 \\
\vdots \\
a_r \\
\gamma_r
\end{array}
\right],
$$

which defines γ_r.

The iteration must update $\mathbf{f}^{(r)}$ to make γ_r equal a_{r+1}. Let $\mathbf{f}^{(r+1)}$ be given by

$$
\begin{bmatrix} f_0^{(r+1)} \\ f_1^{(r+1)} \\ \vdots \\ f_{r-1}^{(r+1)} \\ f_r^{(r+1)} \end{bmatrix}
=
\begin{bmatrix} f_0^{(r)} \\ f_1^{(r)} \\ \vdots \\ f_{r-1}^{(r)} \\ 0 \end{bmatrix}
+ k_r
\begin{bmatrix} f_{r-1}^{(r)} \\ f_{r-2}^{(r)} \\ \vdots \\ f_0^{(r)} \\ 0 \end{bmatrix}
+ \beta_r
\begin{bmatrix} 0 \\ 0 \\ \vdots \\ 0 \\ 1 \end{bmatrix}.
$$

If we can choose k_r and β_r so as to perpetuate the desired form, we have a good algorithm. But for some γ_r and γ_r'

$$
\begin{bmatrix} a_0 & \cdots & a_{r-1} & | & a_r \\ \vdots & & \vdots & | & \vdots \\ a_{r-1} & \cdots & a_0 & | & a_1 \\ \hline a_r & \cdots & a_1 & | & a_0 \end{bmatrix}
\begin{bmatrix} f_0^{(r+1)} \\ \vdots \\ f_{r-1}^{(r+1)} \\ f_r^{(r+1)} \end{bmatrix}
= -
\begin{bmatrix} a_1 \\ \vdots \\ a_r \\ \gamma_r \end{bmatrix}
- k_r
\begin{bmatrix} a_r \\ \vdots \\ a_1 \\ \gamma_r' \end{bmatrix}
$$

$$
+ \beta_r
\begin{bmatrix} a_r \\ \vdots \\ a_1 \\ a_0 \end{bmatrix}
$$

$$
= -
\begin{bmatrix} a_1 \\ \vdots \\ a_r \\ a_{r+1} \end{bmatrix}
$$

provided we choose k_r and β_r so that

$$
k_r - \beta_r = 0
$$

and

$$
-\gamma_r - k_r \gamma_r' + \beta_r a_0 = -a_{r+1},
$$

where

$$
\gamma_r = -\sum_{i=1}^{r} f_{r-i}^{(r)} a_i,
$$

$$
\gamma_r' = -\sum_{i=1}^{r} f_{i-1}^{(r)} a_i.
$$

Hence we must choose k_r and β_r by

$$
k_r = \beta_r,
$$

$$
\beta_r = -\frac{a_{r+1} - \gamma_r}{a_0 - \gamma_r'}.
$$

With these equations, a solution at the $(r-1)$th iteration is propagated into a solution for the rth iteration. Hence, the trivial solution at the zeroth iteration can be recursively propagated into a solution at the nth iteration.

7

Fast Algorithms for the Fourier Transform

A fast Fourier transform (FFT) algorithm is an efficient and nonobvious procedure for computing the (discrete) Fourier transform that is considerably more efficient than is the obvious form of the computation. Notice that the terminology carefully distinguishes between the Fourier transform as a function and the fast Fourier transform as a computational procedure.

There are many FFT algorithms now known, and these can be surprisingly subtle in their construction. The principal FFT algorithms are the Cooley–Tukey FFT, the Good–Thomas FFT, and the Winograd FFT. These algorithms can be mixed and matched in a variety of ways to create an immense library of routines for computing the discrete Fourier transform for many values of the blocklength n.

The most popular blocklengths for the discrete Fourier transform are 256, 512, 1024, 4096, or, in general, those of the form $n = 2^m$ for some positive integer m. This is because these can be computed by the radix-2 Cooley–Tukey FFT, which is easy to use and widely studied. An unfortunate myth that can be attributed to the Cooley–Tukey FFT is the notion that an FFT is practical only if the blocklength is a power of 2. In fact, however, there are good FFT algorithms for almost any blocklength.

7.1 The Cooley–Tukey FFT

The Cooley–Tukey FFT can be used to compute the Fourier transform

$$V_k = \sum_{i=0}^{n-1} \omega^{ik} v_i, \qquad k = 0, \ldots, n-1,$$

whenever the blocklength n is not a prime. To derive the Cooley–Tukey FFT algorithm, suppose that $n = n'n''$. Replace the indices in the Fourier transform by a coarse index and a vernier index as follows:

$$i = i' + n'i'', \qquad i' = 0, \ldots, n' - 1, \quad i'' = 0, \ldots, n'' - 1,$$
$$k = n''k' + k'', \qquad k' = 0, \ldots, n' - 1, \quad k'' = 0, \ldots, n'' - 1.$$

Notice that the structure is different for i and k. Next define the two-dimensional time- and frequency-domain arrays, also called $\mathbf{v}$ and $\mathbf{V}$, given by

$$
\begin{aligned}
v_{i'i''} &= v_{i'+n'i''}, & i' &= 0,\ldots,n'-1, & i'' &= 0,\ldots,n''-1, \\
V_{k'k''} &= V_{n''k'+k''}, & k' &= 0,\ldots,n'-1, & k'' &= 0,\ldots,n''-1.
\end{aligned}
$$

With these definitions, the input and output data vectors are mapped into two-dimensional arrays. Then the Fourier transform becomes

$$
V_{k'k''} = \sum_{i''=0}^{n''-1} \sum_{i'=0}^{n'-1} \omega^{(i'+n'i'')(n''k'+k'')} v_{i'i''}.
$$

Expand the product in the exponent and let $\omega^{n'} = \gamma$ and $\omega^{n''} = \beta$. Because ω has order $n'n''$, the term $\omega^{n'n''i''k'}$ equals 1 and so can be dropped. Then

$$
V_{k'k''} = \sum_{i'=0}^{n'-1} \beta^{i'k'} \left[\omega^{i'k''} \sum_{i''=0}^{n''-1} \gamma^{i''k''} v_{i'i''} \right].
$$

The computation now has the form of an n''-point Fourier transform of every row, followed by componentwise multiplication by $\omega^{i'k''}$, followed by an n'-point Fourier transform of every column.

The number of multiplications and additions now is much less than in the original form of the Fourier transform. At most $n(n' + n'' + 1)$ multiplications in the field F and $n(n' + n'' - 2)$ additions in the field F are required compared to about n^2 multiplications and additions previously. Moreover, if n' or n'' is composite then the Cooley–Tukey FFT can be used to decompose that n'_i-point or that n''-point Fourier transform.

Observe that the components of the Fourier transform are found arranged differently in the array than the components of the signal. If the input data **v** is read into its two-dimensional array by columns, then the output transform **V** will be found stored in its two-dimensional array by rows. Because an array may be stored in one-dimensional physical memory by stacking columns, consecutive components of **V** will not be found in consecutive memory locations. This is known as *address shuffling*. Some provision is needed to return **V** to its natural order.

7.2   Radix-2 Transforms

A blocklength of the form $n = 2^m$ can be factored as $2(2^{m-1})$ or as $(2^{m-1})2$ to form an FFT. Each of these leads to an FFT that is called a *radix-2* Cooley–Tukey FFT. The radix-2 FFT is implemented recursively by breaking down the 2^{m-1}-point Fourier transform in the same way as the 2^m-point Fourier transform. The computation then takes the form of $\log_2 n$ iterations. An iteration is built out of two-point Fourier transforms, each of which requires two additions and no multiplications. Alternating with the two-point Fourier transforms is a multiplication of each of the n components by a power of ω, a total of n multiplications per iteration.

The radix-2 FFT uses $\left(\frac{n}{2}\right)\log_2 n$ multiplications and $n\log_2 n$ additions in the field of the computation. Some of the multiplications may be trivial as when they are multiplications by 1 or by j. They can be bypassed, but at the expense of some irregularity in the organization of the computation.

Likewise, a blocklength of the form $n = 4^m$ can be factored as $4(4^{m-1})$ or as $(4^{m-1})4$ to produce a radix-4 Cooley–Tukey FFT. The computation takes the form of $\log_4 n$ iterations; each iteration consists of 4^{m-1} four-point Fourier transforms followed by the multiplication of each of n components by a power of ω. The four-point Fourier transform in the complex field can be computed with multiplications only by ± 1 and $\pm j$.

The radix-2 (or radix-4) Cooley–Tukey FFT is valid in any field. For example, in the cyclotomic extension of the rationals $\mathbf{Q}^m$ it can be used whenever n is a power of 2 to compute the Fourier transform

$$V_k(x) = \sum_{i=0}^{n-1} x^{ik} v_i(x) \qquad (\bmod\ p(x)).$$

Then it uses $n\log_2 n$ polynomial additions and $\left(\frac{n}{2}\right)\log_2 n$ multiplications — each multiplication is a multiplication (modulo $p(x)$) of a polynomial by a power of x, which consists of only indexing and polynomial additions.

The radix-2 Cooley–Tukey FFT can be used in the Galois field $GF(p)$ whenever that field has a Fourier transform whose blocklength is a power of 2. This occurs whenever $GF(p)$ contains an element whose order is a power of 2.

When p is a Fermat prime $2^m + 1$, every factor of $p - 1$ is a power of 2. The Fourier transform

$$V_k = \sum_{i=0}^{n-1} \omega^{ik} v_i, \qquad k = 0, \ldots, n-1$$

exists whenever n is a divisor of 2^m and ω is an element of order n. Thus the field $GF(2^{16} + 1)$ has Fourier transforms of sizes 2^{16}, 2^{15}, 2^{14}, ..., 2^2, 2. Using the Cooley–Tukey FFT algorithm, a Fourier transform over $GF(2^m + 1)$ can be broken down into a sequence of two-point transforms which can be implemented rather neatly using only $\left(\frac{n}{2}\right)\log_2 n$ multiplications and $\left(\frac{n}{2}\right)\log_2 n$ additions.

The Fourier transforms over $GF(2^{16} + 1)$ with blocklength 32 or less are actually much simpler. This is because the element 2 has order 32. To see this, notice that $2^{16} + 1 = 0$ in $GF(2^{16} + 1)$. Hence $2^{16} = -1$ and $2^{32} = 1$. In general, the element 2 has order $2m$ in $GF(2^m + 1)$ and so can be used as the kernel of a Fourier transform of blocklength $2m$.

The Fourier transform of blocklength 32 in $GF(2^{16} + 1)$ is

$$V_k = \sum_{i=0}^{31} 2^{ik} v_i, \qquad k = 0, \dots, 31$$

and the multiply operation is actually a shift in a binary arithmetic system because it is a multiplication by a power of 2. This Fourier transform is easy to compute because it uses no multiplications. However, a blocklength of 32 is too short for many applications. For larger Fourier transforms it is not possible to replace every multiplication by a simple shift. However, even then there are a variety of methods to minimize the number of multiplications by choice of ω. It is possible to replace nearly all multiplications by shifts, to replace all multiplications by pairs of shifts, or to replace multiplications by shifts of complex integers. We shall illustrate each of these methods in turn.

For example, consider the 1024-point transform in $GF(2^{16} + 1)$

$$V_k = \sum_{i=0}^{1023} \omega^{ik} v_i, \qquad k = 0, \dots, n-1,$$

where ω is now an element of order 1024. We want an ω so that $\omega^{32} = 2$ in $GF(2^{16} + 1)$, and $\omega = 56042 = -9495$ satisfies this requirement. The Cooley–Tukey radix-32 FFT puts the transform in the form

$$V_{n''k'+k''} = \sum_{i'=0}^{31} 2^{i'k'} \left[\omega^{i'k''} \sum_{i''=0}^{31} 2^{i''k''} v_{i'+n'i''} \right].$$

The inner sum is a 32-point Fourier transform for each value of i' and the outer sum is a 32-point Fourier transform for each value of i''. As we have already seen, a 32-point Fourier transform can be computed by a radix-2 Cooley–Tukey FFT with only shifts and adds. The multiplication by the adjustment term $\omega^{i'k''}$ is a nontrivial multiplication, but there are only 1024 such multiplications and they are integer multiplications because ω is the integer 56042. In fact, when i' or k'' equals zero that multiplication is trivial; there are only 931 nontrivial multiplications (and even some of those are shifts). The structure of this FFT is shown in Figure 7.1. In general, a Fourier transform in $GF(2^{16} + 1)$ can be computed using about $m(\lceil \log_{32} m \rceil - 1)$ multiplications in $GF(2^{16} + 1)$, $\left(\frac{n}{2}\right) \log_2 n$ additions in $GF(2^{16} + 1)$, and $\left(\frac{n}{2}\right) \log_2 n$ shifts.

Another technique is to use $\sqrt{2}$ as the kernel to get a Fourier transform of blocklength 64. In $GF(2^{16} + 1)$ we can easily verify that $\sqrt{2} = 2^{12} - 2^4$ by calculating the square if we recall that $2^{16} = -1$. The Fourier transform in $GF(2^{16} + 1)$ then has the appearance

$$V_k = \sum_{i=0}^{63} (\sqrt{2})^{ik} v_i, \qquad k = 0, \dots, 63.$$

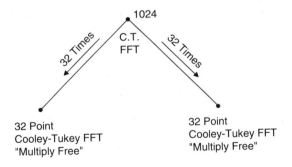

FIGURE 7.1. Structure of a 1024-point FFT in $GF(2^{16} + 1)$.

When the Fourier transform is computed as a radix-2 Cooley–Tukey FFT, all multiplications are by constants of the form $2^a \pm 2^b$, which can be implemented as a pair of shifts and an addition. Furthermore all even powers of $\sqrt{2}$ are powers of 2, which are simple shifts, and the Cooley–Tukey FFT uses mostly even powers of ω.

For a third example, let the elements of $GF((2^{16} + 1)^2)$ be written as $a + jb$, where j is a zero of $x^2 + x + 1$. The element j has order 3 because $j^2 = -j - 1$ and $j^3 = -j^2 - j = 1$. The element 2 has order 32 just as it did in $GF(2^{16} + 1)$. Consequently, because 32 and 3 are relatively prime, the element $2j$ must have order 96. Thus in $GF((2^{16} + 1)^2)$ we have the following Fourier transform of blocklength 96:

$$V_k = \sum_{i=0}^{95} (2j)^{ik} v_i, \qquad k = 0, \dots, 95,$$

which can be used to compute cyclic convolutions of blocklength 96. The powers of $2j$ can be written $(2j)^{ik} = 2^{ik} j^{ik}$ and powers of j modulo $j^2 + j + 1$ follow the cycle $j, -j - 1, 1$. Therefore $(2j)^{ik}$ is always a constant of the form $\pm 2^a j$ or $\pm 2^a (j + 1)$ for some a and multiplication by such a constant is little more than a shift. The Fourier transforms can be computed with no multiplications.

7.3 The Good–Thomas FFT

When the discrete Fourier transform has blocklength n that is not a power of 2 (or a power of any prime), then another FFT known as the Good–Thomas FFT is a little more efficient than the Cooley–Tukey FFT as measured by the number of multiplications and additions. The Good–Thomas FFT is based on the Chinese remainder theorem for integers and will break a Fourier transform of blocklength $n'n''$ into a two-dimensional $n' \times n''$

Fourier transform provided that n' and n'' are relatively prime.

To derive the Good–Thomas FFT algorithm in the field F, suppose that $n = n'n''$. Replace the indices in the Fourier transform by a pair of indices as follows:

$$
\begin{aligned}
i' &= i \quad (\bmod\ n'),\\
i'' &= i \quad (\bmod\ n'').
\end{aligned}
$$

This is the map of the input index i down the extended diagonal of a two-dimensional array indexed by (i', i''). The Chinese remainder theorem asserts that there exist integers N' and N'' such that

$$
N'n' + N''n'' = 1.
$$

Then the input index can be recovered as follows:

$$
i = i'N''n'' + i''N'n' \quad (\bmod\ n).
$$

The output index is described somewhat differently. Define

$$
\begin{aligned}
k' &= N''k \quad (\bmod\ n'),\\
k'' &= N'k \quad (\bmod\ n'').
\end{aligned}
$$

The output index k can be recovered as follows:

$$
k = n''k' + n'k'' \quad (\bmod\ n).
$$

To verify this, write it out.

$$
\begin{aligned}
k &= n''(N''k + Q_1 n') + n'(N'k + Q_2 n'') \quad &(\bmod\ n'n'')\\
&= k(n''N'' + n'N') \quad &(\bmod\ n)\\
&= k
\end{aligned}
$$

because $n''N'' + n'N' = 1$. Now, with these new indices, the discrete Fourier transform becomes

$$
V_{n''k' + n'k''} = \sum_{i''=0}^{n''-1} \sum_{i'=0}^{n'-1} \omega^{(i'N''n'' + i''N'n')(n''k' + n'k'')} v_{i'N''n'' + i''N'n'}.
$$

Multiply out the exponent. Terms in the exponent involving $n'n''$ can be dropped because ω has order $n'n''$. The input and output vectors are now viewed as two-dimensional arrays. Then

$$
\begin{aligned}
V_{k'k''} &= \sum_{i'=0}^{n'-1} \sum_{i''=0}^{n''-1} \omega^{N''(n'')^2 i'k'} \omega^{N'(n')^2 i''k''} v_{i'i''}\\
&= \sum_{i'=0}^{n'-1} \beta^{i'k'} \sum_{i''=0}^{n''-1} \gamma^{i''k''} v_{i'i''},
\end{aligned}
$$

where $\beta = \omega^{N''(n'')^2}$ and $\gamma = \omega^{N'(n')^2}$. The terms β and γ are the roots of unity of order n' and n'', respectively, in the field F that are needed for the n'-point Fourier transforms and the n''-point Fourier transforms, respectively. To see this for β, notice that $\beta = (\omega^{n''})^{N''n''}$. Let $\omega^{n''} = \omega'$, which is the primary n'th root of unity. Then because $N''n'' = 1 - N'n'$ and $(\omega')^{n'} = 1$, it follows that $\beta = \omega'$.

The equation is now in the form of a two-dimensional $n' \times n''$-point Fourier transform. The number of multiplications is about $n(n' + n'')$ in the field of the computation, and the number of additions is about the same. The Fourier transform on the rows or on the columns, if the size is composite, can in turn be simplified by another application of the fast Fourier transform. In this way, a transform whose blocklength n has relatively prime factors n_ℓ can be broken down into a form requiring about $n \sum_\ell n_\ell$ multiplications and additions. Notice that to derive this FFT we have only manipulated indices; the rearrangement makes no use of the arithmetic of the field F in which the Fourier transform is given.

One can choose either the Cooley–Tukey or the Good–Thomas FFT to compute a Fourier transform. It is even possible to build a hybrid Fourier transform algorithm using both the Cooley–Tukey and the Good–Thomas FFT. For example, a 63-point Fourier transform can be broken into a seven-point Fourier transform and a nine-point Fourier transform using the Good–Thomas FFT; the nine-point transform can then be broken into two three-point transforms using the Cooley–Tukey FFT.

7.4   FFT Algorithms for Subblocks

One may wish to compute only a portion of the vector of spectral components. We shall study the case in which only the first n' components of the Fourier transform are to be computed, where $n = n'n''$ and the factors n' and n'' are relatively prime. That is, the task is to compute

$$V_k = \sum_{i=0}^{n-1} \omega^{ik} v_i$$

for $k = 0, \ldots, n' - 1$. The algorithm that will be developed for this task resembles the Good–Thomas FFT in the way it uses the Chinese remainder theorem. Let

$$
\begin{aligned}
i' &= N''i \quad (\mathrm{mod}\ n'), \\
i'' &= N'i \quad (\mathrm{mod}\ n''),
\end{aligned}
$$

where N' and N'' are the unique integers satisfying

$$N'n' + N''n'' = 1.$$

Then
$$i = n''i' + n'i'' \qquad (\mathrm{mod}\ n)$$

and we can write

$$
\begin{aligned}
V_k &= \sum_{i'=0}^{n'-1} \sum_{i''=0}^{n''-1} \omega^{(n''i'+n'i'')k} v_{n''i'+n'i''} \\
&= \sum_{i''=0}^{n''-1} \gamma^{i''k} \sum_{i'=0}^{n'-1} \beta^{i'k} v_{i'i''}, \qquad k = 0, \ldots, n'-1,
\end{aligned}
$$

where $\beta = \omega^{n''}$, $\gamma = \omega^{n'}$ and

$$v_{i'i''} = v_{n''i'+n'i''}.$$

The two-dimensional array $\{v_{i'i''}\}$ can be converted back into a one-dimensional array by stacking columns. This vector, which we also call $\mathbf{v}$, is a permutation of the components of the original $\mathbf{v}$. Then the n' components of $\mathbf{V}$ can be re-expressed as a matrix-vector product

$$\mathbf{V} = [\mathbf{W} \quad \mathbf{DW} \quad \mathbf{D}^2\mathbf{W} \quad \mathbf{D}^3\mathbf{W} \ \ldots \ \mathbf{D}^{n''-1}\mathbf{W}]\mathbf{v},$$

where $\mathbf{V} = (V_0, V_1, \ldots, V_{n-1})^T$ is a spectral vector of blocklength n', and $\mathbf{v}$ is the permuted input vector of blocklength n. The matrix $\mathbf{W}$ is the Fourier transform matrix

$$
\mathbf{W} = \begin{bmatrix}
1 & 1 & 1 & \cdots & 1 \\
1 & \beta & \beta^2 & \cdots & \beta^{n'-1} \\
\vdots & & & & \\
1 & \beta^{n'-1} & \beta^{n'-2} & \cdots & \beta
\end{bmatrix},
$$

and the matrix $\mathbf{D}$ is the diagonal matrix

$$
\mathbf{D} = \begin{bmatrix}
1 & & & & 0 \\
& \gamma & & & \\
& & \gamma^2 & & \\
& & & \ddots & \\
0 & & & & \gamma^{n'-1}
\end{bmatrix}.
$$

The computation of $\mathbf{V}$ is organized as follows. Form the two-dimensional array with elements

$$v_{i'i''} = v_{N''i(\mathrm{mod}\ n'),N'i(\mathrm{mod}\ n'')}.$$

Multiply each column by $\mathbf{W}$. Then multiply the ℓth column by $\mathbf{D}^\ell$ for $\ell = 0, \ldots, n'' - 1$ and sum the columns. The resulting n'-dimensional vector consists of the first n' components of the spectrum. Multiplication by

$\mathbf{D}^\ell$ and summing can be organized conveniently using *Horner's rule* for evaluating a polynomial

$$\sum_{i=0}^{n-1} a_i x^i = x(((\ldots x(x(a_{n-1}x + a_{n-2}) + a_{n-3}) + \ldots + a_1) + a_0.$$

An alternative formulation is to first compute $\mathbf{W}^{-1}\mathbf{V}$ which is then multiplied by $\mathbf{W}$ to compute $\mathbf{V}$. Multiplying the earlier equation by $\mathbf{W}^{-1}$ gives

$$\begin{aligned}\mathbf{W}^{-1}\mathbf{V} &= [\mathbf{I} \quad \mathbf{W}^{-1}\mathbf{DW} \quad \mathbf{W}^{-1}\mathbf{D}^2\mathbf{W} \ \ldots \ \mathbf{W}^{-1}\mathbf{D}^{n''-1}\mathbf{W}]\mathbf{v} \\ &= [\mathbf{I} \quad \mathbf{A} \quad \mathbf{A}^2 \quad \mathbf{A}^3 \ \ldots \ \mathbf{A}^{n''-1}]\mathbf{v},\end{aligned}$$

where $\mathbf{A} = \mathbf{W}^{-1}\mathbf{DW}$ is a circulant matrix because, by the convolution theorem, it corresponds to a convolution. Using Horner's rule, this can be put in the form

$$\mathbf{W}^{-1}\mathbf{V} = \mathbf{A}(((\ldots \mathbf{A}(\mathbf{A}(\mathbf{A}\mathbf{v}_{n-1} + \mathbf{v}_{n-2}) + \mathbf{v}_{n-3}) + \ldots + \mathbf{v}_1) + \mathbf{v}_0.$$

The computation proceeds in the time domain by alternately adding an n'-vector of input components, then convolving with a row of $\mathbf{A}$. When this is completed, multiplication by $\mathbf{W}$ gives $\mathbf{V}$.

Essentially the same ideas can be used for interpolation to compute a block of n points from n' known points if it is also known that only the first n' components of the spectrum are nonzero. The method only requires that $n = n'n''$, where n' and n'' are relatively prime.

The inverse Fourier transform for this problem is written

$$v_i = \frac{1}{n}\sum_{k=0}^{n'-1} \omega^{-ik}V_k, \qquad i = 0,\ldots,n-1.$$

Using the two-dimensional rearrangement of the Good–Thomas algorithm, this becomes

$$v_{i'i''} = \frac{1}{n}\sum_{k=0}^{n'-1} \beta^{i'k}\gamma^{i''k}V_k.$$

For a fixed value of i'' this has the form of a Fourier transform of a vector with components $\gamma^{i''k}V_k$ for $k = 0,\ldots,n'-1$. This can be represented as a diagonal matrix multiplying $\mathbf{V}$, where the matrix has diagonal entries $\gamma^{i''k}$. Therefore, by stacking columns of the two-dimensional array, the computation becomes

$$\begin{bmatrix} \mathbf{v}_0 \\ \mathbf{v}_1 \\ \mathbf{v}_2 \\ \vdots \\ \mathbf{v}_{n''-1} \end{bmatrix} = \frac{1}{n}\begin{bmatrix} \mathbf{W} \\ \mathbf{WD} \\ \mathbf{WD}^2 \\ \vdots \\ \mathbf{WD}^{n''-1} \end{bmatrix}[\mathbf{V}] = \begin{bmatrix} \mathbf{W} \\ \mathbf{WD} \\ \mathbf{WD}^2 \\ \vdots \\ \mathbf{WD}^{n''-1} \end{bmatrix}[\mathbf{W}^{-1}\mathbf{v}_0].$$

Therefore

$$
\begin{bmatrix}
\mathbf{v}_0 \\
\mathbf{v}_1 \\
\mathbf{v}_2 \\
\vdots \\
\mathbf{v}_{n''-1}
\end{bmatrix}
=
\begin{bmatrix}
\mathbf{I} \\
\mathbf{A} \\
\mathbf{A}^2 \\
\vdots \\
\mathbf{A}^{n''-1}
\end{bmatrix}
[\mathbf{v}_0],
$$

which leads to the recursion

$$
\mathbf{v}_{\ell+1} = \mathbf{A}\mathbf{v}_\ell,
$$

where

$$
\mathbf{A} = \mathbf{W}\mathbf{D}\mathbf{W}^{-1}.
$$

Each time ℓ is incremented, this recursion interpolates a set of n' equally spaced points to produce a batch of n' points.

7.5   FFT Algorithms Using Algebraic Integers

To compute a Fourier transform in the complex field, the complex root of unity ω must be quantized and the data must also be quantized. The conventional method of quantization is to scale ω and then round to the nearest Gaussian integer. A larger scale factor leads to better accuracy in the computation but also means that a longer wordlength must be carried in the computations.

An alternative approach is to approximate ω and the data by elements of the ring of algebraic integers $\mathbf{Z}[\omega']$, where ω' is a 2^ν-root of unity. In particular, $\mathbf{Z}[e^{j2\pi/8}]$ and $\mathbf{Z}[e^{j2\pi/16}]$ are dense in the complex plane; either is a suitable choice for a ring of algebraic integers into which a complex computation can be embedded.

Figure 7.2 gives approximations (computed by Cozzens and Finkelstein) in $\mathbf{Z}[e^{j2\pi/16}]$ to ξ^k, where $\xi = e^{-j2\pi/256}$. The approximations

$$
\widetilde{\xi^k} = \sum_{j=0}^{7} a_j \omega^j
$$

were made so that the magnitude of the error $|\xi^k - \widetilde{\xi^k}|$ is at most $1/512$ (equivalent to nine bits of accuracy) and the integer coefficients a_j are as small as possible.

The Fourier transform of blocklength 256,

$$
V_k = \sum_{i=0}^{255} \xi^{ik} v_i,
$$

k	a_0	a_1	a_2	a_3	a_4	a_5	a_6	a_7	Error
0	1	0	0	0	0	0	0	0	0.0
1	3	-3	5	-1	1	-4	-1	5	0.0011
2	1	4	-1	-2	2	-2	-1	4	0.0014
3	1	-4	5	-3	2	-1	0	-1	0.0012
4	0	-5	2	4	1	-2	-4	1	0.0009
5	3	-3	1	0	2	0	-4	3	0.0011
6	0	-1	3	-1	3	-4	-2	3	0.0011
7	3	2	2	-3	-1	-2	3	3	0.0012
8	0	1	-2	4	-4	2	-1	0	0.0014
9	2	-3	-3	3	3	-1	-2	-1	0.0006
10	4	-3	-1	4	-1	-2	0	2	0.0010
11	-3	4	0	-2	0	-1	3	-3	0.0011
12	-4	-5	4	3	0	0	-4	-3	0.0005
13	-2	0	4	4	-5	-1	-4	5	0.0018
14	1	-1	2	0	-3	4	-4	2	0.0015
15	2	-5	-1	4	4	-4	-3	1	0.0009

FIGURE 7.2. Approximations of ξ^k in $\mathbf{Z}[\omega]$.

is now approximated by

$$V_k(\omega) = \sum_{i=0}^{255} \left[\sum_{j=0}^{7} a_j \omega^j \right] v_i(\omega),$$

where $v_i(\omega)$ is an (approximate) representation of v_i and $V_k(\omega)$ is an (approximate) representation of V_k. All of the coefficients are small integers, so that the general multiplications of real numbers have been replaced by multiplications of polynomials in the variable ω that have small integer coefficients.

7.6   The Winograd FFT

The *Winograd small FFT* is a way of constructing highly optimized routines for small prime (or prime power) blocklengths — blocklengths of 2, 3, 4, 5, 7, 9, and 16 are typical. The *Winograd large FFT* is a method of binding these small pieces together to get a large FFT. For the Fourier transform of a Reed–Solomon code of blocklength 255, for example, one must use Winograd small FFT algorithms of blocklengths 3, 5, and 17, and bind these with the Winograd large FFT into a FFT of blocklength 255. We shall discuss only the structure of the Winograd small FFT.

Just as for the Cooley–Tukey and Good–Thomas FFTs, the Winograd FFT can be used in any field including a finite field. However, unlike those FFTs, the Winograd FFT does more than just permute components by rearranging indices. The structure of the FFT does depend on the structure of the field. In principle, the Winograd small FFT has to be rederived for each field of interest, though often only minor adjustments will suffice to move the Winograd small FFT from one field to another.

The Winograd small FFT is based in part on the *Rader algorithm*, which turns a Fourier transform of prime blocklength p into a cyclic convolution of blocklength $p - 1$. Over any field F, let

$$\mathbf{v} = (v_0, v_1, \ldots, v_{p-1})$$

be a vector of blocklength p, where p is a prime. The Fourier transform

$$V_k = \sum_{i=0}^{p-1} \omega^{ik} v_i, \qquad k = 0, \ldots, p - 1$$

can be rewritten with the indices $k = 0$ and $i = 0$ treated separately as

$$V_0 = \sum_{i=0}^{p-1} v_i,$$

$$V_k = v_0 + \sum_{i=1}^{p-1} \omega^{ik} v_i, \qquad k = 1, \ldots, p - 1.$$

Now introduce the Rader permutation (which was already used in Section 3.8). Let π be a prime in $GF(p)$ and let $i = \pi^r$ and $k = \pi^{-s}$. Then

$$V_{\pi^{-s}} = v_0 + \sum_{i=1}^{p-1} \omega^{\pi^r \pi^{-s}} v_{\pi^r}.$$

Let $V'_s = V_{\pi^{-s}} - v_0$ and $v'_r = v_{\pi^r}$ and let

$$g(x) = \sum_{r=0}^{p-2} \omega^{\pi^{-r}} x^r.$$

The equation now takes the form of a cyclic convolution

$$V'(x) = g(x) v'(x) \qquad (\text{mod } x^{p-1} - 1),$$

where $V'(x)$ and $v'(x)$ are the polynomials with coefficients V'_s and v'_r, respectively.

The construction of the Winograd FFT is completed by using the Winograd cyclic convolution to compute the cyclic convolution. This requires that $x^{p-1} - 1$ be factored into its irreducible factors over F. The polynomials $g(x)$ and $v'(x)$ are represented by their residues modulo the irreducible factors, the residues are multiplied, and $V'(x)$ is recovered from these products by using the Chinese remainder theorem. Because the factorization of $x^{p-1} - 1$ depends on the field F, the Winograd FFT depends on the field F.

The Winograd FFT is the most efficient FFT as measured by the number of multiplications, but it is also the most difficult to understand. It does not allow tight programming loops, so the instruction memory may be larger than with other methods. The running time, however, may be less. The Winograd FFT must be programmed with careful attention to scaling or else computational noise will be larger than with other methods. The Winograd FFT also performs poorly on a processor that is designed specifically for the Cooley–Tukey FFT. These are some of the reasons why the Winograd FFT has been slow to find acceptance. However, the Winograd FFT does reduce considerably the number of multiplications needed, and also reduces the number of additions by a small amount.

8

Decoding of Cyclic Codes

Cyclic codes have been defined from a spectral point of view based on the Fourier transform. Toeplitz matrices are directly related to spectral analysis, so it will not be surprising that the algorithms for inverting Toeplitz systems of equations can be applied to the decoding of cyclic codes. This chapter will make this connection explicit by developing a number of decoding algorithms for Reed–Solomon codes.

8.1   Decoding of Reed–Solomon Codes

The spectral estimation problem that arises in the decoding of Reed–Solomon codes is the following: Find the vector $\mathbf{e}$ with the smallest number of nonzero components such that the Fourier transform $\mathbf{E}$ has $2t$ components E_k for $k = 0, \ldots, 2t - 1$ that agree with the known values of these $2t$ components. This is the most likely transmitted codeword if the channel has a probability of symbol error smaller than one half. The procedure that we shall derive is predicated on the assumption that there is a solution $\mathbf{e}$ with at most t nonzero components because this is the error-correcting capability of the code.

The codeword $\mathbf{c}$ is transmitted and the channel makes errors described by the vector $\mathbf{e}$, which is nonzero in not more than t places. The received word $\mathbf{v}$ is written componentwise as

$$v_i = c_i + e_i, \qquad i = 0, \ldots, n - 1.$$

The decoder must process the received word $\mathbf{v}$ so as to remove the error word $\mathbf{e}$; the data is then recovered from the corrected codeword $\mathbf{c}$. The received noisy codeword $\mathbf{v}$ has a Fourier transform

$$V_k = \sum_{i=0}^{n-1} \omega^{ik} v_i, \qquad k = 0, \ldots, n - 1$$

with components $V_k = C_k + E_k$ for $k = 0, \ldots, n - 1$. But, by construction of a Reed–Solomon code,

$$C_k = 0, \qquad k = 0, \ldots, 2t - 1.$$

Hence
$$V_k = E_k, \qquad k = 0, \ldots, 2t - 1.$$

The block of $2t$ components of $\mathbf{v}$ gives us a window through which we can look at $2t$ consecutive components of the n components of the transform of the error pattern. These $2t$ components of $\mathbf{E}$ are referred to as the *syndromes* of $\mathbf{E}$. The decoder must find all n components of $\mathbf{E}$ given a segment consisting of $2t$ consecutive components of $\mathbf{E}$ and the additional condition that at most t components of the time domain error pattern $\mathbf{e}$ are nonzero. Once $\mathbf{E}$ is known, the computation is trivial because $C_k = V_k - E_k$. From $\mathbf{C}$ one can compute $\mathbf{c}$. The data symbols are recovered easily in various ways depending on the method of encoding.

We will use the properties of the Fourier transform to find a procedure to compute $\mathbf{E}$. Because $\mathbf{e}$ has weight at most t, we can find a locator polynomial $\Lambda(x)$ of degree at most t and with $\Lambda_0 = 1$ such that

$$\Lambda(x)E(x) = 0 \qquad (\bmod\ x^n - 1),$$

where

$$E(x) = \sum_{k=0}^{n-1} E_k x^k.$$

We can rewrite the convolution as the equation of an autoregressive filter with t taps:

$$E_j = -\sum_{k=1}^{t} \Lambda_k E_{j-k}, \qquad j = 0, \ldots, n - 1.$$

If the taps of the filter were known, then the remaining components of $\mathbf{E}$ can be found by shifting the autoregressive filter as described by

$$E_j = -\sum_{k=1}^{t} \Lambda_k E_{j-k}, \qquad j = 2t, \ldots, n - 1.$$

But we know $2t$ components of $\mathbf{E}$. The t equations

$$E_j = -\sum_{k=1}^{t} \Lambda_k E_{j-k}, \qquad j = t, \ldots, 2t - 1$$

involve only known components of $\mathbf{E}$. These t equations are linear in the unknown components Λ_k. Hence we can find $\mathbf{\Lambda}$ by solving this system of linear equations. This system of equations has come up repeatedly and we need only repeat some of the main ideas in the context of the decoding problem.

The locator polynomial $\Lambda(x)$ has the form

$$\Lambda(x) = \prod_{\ell=1}^{\nu} (1 - x\omega^{i_\ell}),$$

where i_ℓ for $\ell = 1, \ldots, \nu \le t$ are the locations of the nonzero e_i. The inverse Fourier transform

$$\lambda_i = \frac{1}{n} \sum_{k=0}^{n-1} \Lambda_k \omega^{-ik}$$

can be obtained from $\Lambda(x)$ by evaluating $\Lambda(x)$ at $x = \omega^{-i}$. That is,

$$\lambda_i = \frac{1}{n} \Lambda(\omega^{-i}).$$

Therefore

$$\lambda_i = \frac{1}{n} \prod_{\ell=1}^{\nu} (1 - \omega^{-i} \omega^{i_\ell}).$$

Hence $\lambda_i = 0$ if and only if $e_i \ne 0$. We conclude that, in the time domain, $\lambda_i e_i = 0$ for all i. Therefore, because a product in the time domain corresponds to a convolution in the frequency domain, we see that the convolution in the frequency domain is equal to zero.

$$\boldsymbol{\Lambda} * \mathbf{E} = \mathbf{0}.$$

Hence a polynomial $\Lambda(x)$ that solves the cyclic convolution $\Lambda(x)E(x)$ (mod $x^n - 1$) does exist; it is the error-locator polynomial.

To solve the t equations,

$$\sum_{k=1}^{t} \Lambda_k E_{j-k} = -E_j, \qquad j = t, \ldots, 2t - 1$$

is the computational problem of inverting a Toeplitz system of equations and has been studied in Chapter 6. The Berlekamp–Massey algorithm discussed in Section 6.2 has priority within the literature of decoding Reed–Solomon codes and is regarded as the most efficient algorithm.

8.1.1   THE FORNEY ALGORITHM

Generation of $\mathbf{E}$ by recursive extension using the $2t$ known components of $\mathbf{E}$ and the autoregressive filter with taps $\boldsymbol{\Lambda}$ requires computing $n - 2t$ outputs of the filter; each output requires up to t multiplications. An alternative procedure is the Forney algorithm.

The cyclic convolution of the error polynomial and the locator polynomial can be written in acyclic form as

$$\Lambda(x)E(x) = \Gamma(x) - x^n \Gamma(x),$$

where both terms on the right involve a polynomial $\Gamma(x)$ because the right side is zero modulo $x^n - 1$. Moreover, because of Theorem 3.1.1 we know that $\deg \Gamma(x) < t$.

The formal derivative of this equation is

$$\Lambda'(x)E(x) + \Lambda(x)E'(x) = \Gamma'(x) - nx^{n-1}\Gamma(x) - x^n\Gamma'(x).$$

Set $x = \omega^{-i}$, noting that $\omega^{-n} = 1$. Then

$$\Lambda'(\omega^{-i})E(\omega^{-i}) + \Lambda(\omega^{-i})E'(\omega^{-i}) = -n\omega^i\Gamma(\omega^{-i}).$$

But we know that $e_i = \frac{1}{n}E(\omega^{-i})$ is nonzero only if $\Lambda(\omega^{-i})$ is zero. Therefore the error pattern is given by

$$e_i = \begin{cases} 0, & \text{if } \Lambda(\omega^{-i}) = 0, \\ -\dfrac{\Gamma(\omega^{-i})}{\omega^{-i}\Lambda'(\omega^{-i})}, & \text{if } \Lambda(\omega^{-i}) \neq 0, \end{cases}$$

where $\Lambda'(x)$ is computed as the formal derivative of $\Lambda(x)$,

$$\Lambda'(x) = \sum_{k=0}^{t} k\Lambda_k x^{k-1},$$

and by Theorem 3.1.1, $\Gamma(x)$ is given by $\Lambda(x)E(x) \pmod{x^t}$. It is equivalent (and more convenient) to write this as

$$\Gamma(x) = \Lambda(x)E(x) \qquad \pmod{x^{2t}},$$

because the superfluous coefficients are all zero.

8.1.2   THE BERLEKAMP ALGORITHM

Rather than compute $\Gamma(x)$ after the computation of $\Lambda(x)$ is complete, an iteration for $\Gamma(x)$ can be executed in parallel with each iteration of the Berlekamp–Massey algorithm. The concurrent iteration of $\Lambda(x)$ and $\Gamma(x)$ as in Figure 8.1 is called the *Berlekamp algorithm*.

□ **Theorem 8.1.1 (Berlekamp algorithm)** *If*

$$\begin{bmatrix} \Gamma^{(0)}(x) \\ A^{(0)}(x) \end{bmatrix} = \begin{bmatrix} 0 \\ -x^{-1} \end{bmatrix}$$

and for $r = 1, \ldots, 2t$

$$\begin{bmatrix} \Gamma^{(r)}(x) \\ A^{(r)}(x) \end{bmatrix} = \begin{bmatrix} 1 & -\Delta_r x \\ \delta_r \Delta_r^{-1} & (1 - \delta_r)x \end{bmatrix} \begin{bmatrix} \Gamma^{(r-1)}(x) \\ A^{(r-1)}(x) \end{bmatrix},$$

then $\Gamma^{(2t)}(x) = \Gamma(x)$.

Proof Define the iterates

$$\begin{aligned} \Gamma^{(r)}(x) &= E(x)\Lambda^{(r)}(x), & \pmod{x^r}, \\ A^{(r)}(x) &= E(x)B^{(r)}(x) - x^{r-1}, & \pmod{x^r}, \end{aligned}$$

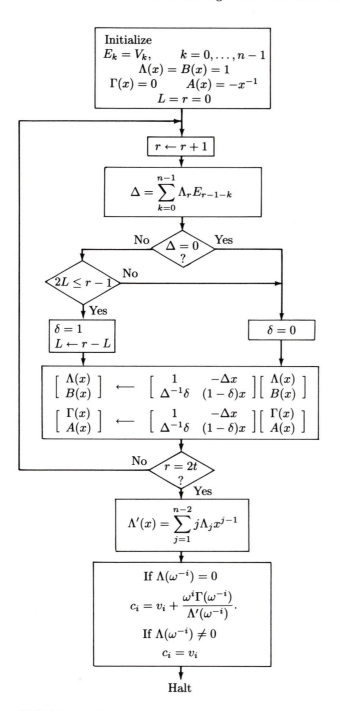

FIGURE 8.1. A decoder that uses the Berlekamp algorithm.

for $r = 1, \ldots, 2t$. Clearly, $\Gamma(x)$ is equal to $\Gamma^{(2t)}(x)$.

In particular, at iteration 1 these start out as

$$
\begin{aligned}
\Gamma^{(1)}(x) &= E(x)(1 - \Delta x), &&(\mathrm{mod}\ x^1) \\
&= E_0
\end{aligned}
$$

and

$$
\begin{aligned}
A^{(1)}(x) &= \begin{cases} E(x)x - x^0 & (\mathrm{mod}\ x^1) & \text{if } E_0 = 0, \\ E(x)E_0^{-1} - x^0 & (\mathrm{mod}\ x^1) & \text{if } E_0 \neq 0, \end{cases} \\
&= \begin{cases} -1 & \text{if } E_0 = 0, \\ 0 & \text{if } E_0 \neq 0. \end{cases}
\end{aligned}
$$

To verify the rth iteration, notice that because

$$
\Gamma_k^{(r-1)} = \sum_{j=0}^{n-1} \Lambda_j^{(r-1)} E_{k-j}, \qquad k = 0, \ldots, r-2
$$

and

$$
\Delta_r = \sum_{j=0}^{n-1} \Lambda_j^{(r-1)} E_{r-1-j}
$$

we have

$$
\Gamma^{(r-1)}(x) = \Delta_r x^{r-1} = E(x) \Lambda^{(r-1)}(x) \qquad (\mathrm{mod}\ x^r).
$$

Using the iteration rule of the Berlekamp–Massey algorithm to expand the right side of

$$
\begin{bmatrix} \Gamma^{(r)}(x) \\ A^{(r)}(x) \end{bmatrix} = \begin{bmatrix} E(x)\Lambda^{(r)}(x) \\ E(x)B^{(r)}(x) - x^{r-1} \end{bmatrix} \qquad (\mathrm{mod}\ x^r)
$$

leads to

$$
\begin{aligned}
\begin{bmatrix} \Gamma^{(r)}(x) \\ A^{(r)}(x) \end{bmatrix} &= \begin{bmatrix} 1 & -\Delta \\ \Delta^{-1}\delta & (1-\delta) \end{bmatrix} \begin{bmatrix} E(x)\Lambda^{(r-1)}(x) \\ xE(x)B^{(r-1)}(x) \end{bmatrix} \\
&\quad - \begin{bmatrix} 0 \\ x^{r-1} \end{bmatrix} \qquad (\mathrm{mod}\ x^r) \\
&= \begin{bmatrix} 1 & -\Delta \\ \Delta^{-1}\delta & (1-\delta) \end{bmatrix} \begin{bmatrix} \Gamma^{(r-1)}(x) + \Delta x^{r-1} \\ xA^{(r-1)}(x) + x^{r-1} \end{bmatrix} - \begin{bmatrix} 0 \\ x^{r-1} \end{bmatrix} \\
&= \begin{bmatrix} 1 & -\Delta \\ \Delta^{-1}\delta & (1-\delta) \end{bmatrix} \begin{bmatrix} \Gamma^{(r-1)}(x) \\ xA^{(r-1)}(x) \end{bmatrix}.
\end{aligned}
$$

This is the desired iteration. The first iteration gives

$$
\begin{bmatrix} \Gamma^{(1)}(x) \\ A^{(1)}(x) \end{bmatrix} = \begin{bmatrix} 1 & -\Delta x \\ \delta\Delta^{-1} & (1-\delta)x \end{bmatrix} \begin{bmatrix} 0 \\ -x^{-1} \end{bmatrix},
$$

which, because $E_0 = \Delta$, reduces to

$$\Gamma^{(0)}(x) = E_0$$

and

$$A^{(0)}(x) = \begin{cases} -1, & E_0 = 0, \\ 0, & E_0 \neq 0, \end{cases}$$

as is required for the first iteration. Thus the initialization provides the right start. □

8.2 Erasures and Errors Decoding

A received word on a channel that makes both errors and erasures consists of channel input symbols, some of which may be in error, and blanks denoting erasures. The decoder must correct the errors and fill the erasures. Any pattern of ν errors and ρ erasures can be decoded provided that

$$d_{\min} \geq 2\nu + 1 + \rho.$$

To decode a received word with ρ erasures, it is necessary to find a codeword that differs from the unerased portion of the received word in the fewest number of places provided the above inequality is satisfied. At most one such solution exists.

Decoding algorithms for Reed–Solomon codes can be extended to algorithms for decoding both erasures and errors. We treat only codes with the pattern of spectral zeros beginning at $k_0 = 0$.

Suppose that ρ erasures are made in locations $i_1, i_2, \ldots, i_\rho$. At these known locations, the received word v_i for $i = 0, \ldots, n-1$ has blanks, which we will initially fill with zeros. Define the erasure vector as that n-vector having component f_{i_ℓ} for $\ell = 1, \ldots, \rho$ equal to the erased symbol, and in other components $f_i = 0$. Then

$$v_i = c_i + e_i + f_i, \qquad i = 0, \ldots, n-1.$$

Define the modified received word as $v_i' = \psi_i v_i$, where ψ is any vector that is zero at every erased location and otherwise nonzero. Then

$$\begin{aligned} v_i' &= \psi_i(c_i + e_i + f_i) = \psi_i c_i + \psi_i e_i \\ &= c_i' + e_i', \end{aligned}$$

where the modified error vector is $e_i' = \psi_i e_i$ and the modified codeword is $c_i' = \psi_i c_i$. The modified error vector $\mathbf{e}'$ is nonzero in those components where $\mathbf{e}$ is nonzero.

The problem now is to decode $\mathbf{v}'$ to find $\mathbf{e}'$, which we know how to do if there are at least 2ν consecutive components where $C_k' = 0$.

The next step in the development is to choose ψ. Let $U_\ell = \omega^{i_\ell}$ for $\ell = 1, \ldots, \rho$ point to the erasure locations. Define the *erasure-locator polynomial* $\Psi(x)$ by

$$\sum_{k=0}^{n-1} \Psi_k x^k = \prod_{\ell=1}^{\rho} (1 - x U_\ell).$$

The inverse Fourier transform of the vector Ψ has components ψ_i equal to zero whenever $f_i \neq 0$. Therefore $\psi_i f_i = 0$ for all i. In the frequency domain,

$$V' = (\Psi * C) + E'.$$

But Ψ is nonzero only in a block of length $\rho+1$, and by the construction of a BCH code C is zero in a block of $2t$ consecutive components. Consequently, the convolution $\Psi * C$ is zero in a block of $2t - \rho$ consecutive components. In this block, define the modified syndrome by

$$\begin{aligned} S_k' &= V_k' \\ &= (\Psi * V)_k = E_k', \end{aligned}$$

which has enough syndrome components to describe the correctable patterns of the modified error vector e'. From these $2t - \rho$ known values of E' we can find the error-locator polynomial $\Lambda(x)$ provided the number of errors ν is not more than $(2t - \rho)/2$.

Once the error-locator polynomial is known, we can combine it with the erasure-locator polynomial and proceed as in the errors-only case. Define the error and erasure locator polynomial

$$\overline{\Lambda}(x) = \Psi(x)\Lambda(x).$$

The inverse Fourier transform of $\overline{\Lambda}$ is zero at every error or erasure. That is, $\overline{\lambda}_i = 0$ if $e_i \neq 0$ or $f_i \neq 0$. Therefore $\overline{\lambda}_i(e_i + f_i) = 0$,

$$\overline{\Lambda} * (E + F) = 0,$$

and $\overline{\Lambda}$ is nonzero in a block of length of at most $\nu + \rho + 1$. Hence the $2t$ known values of $E + F$ can be recursively extended to n values by using this convolution equation and the known value of $\overline{\Lambda}$. Then

$$C_k = V_k - (E_k + F_k).$$

An inverse Fourier transform completes the decoding.

In summary, to compute the error locator polynomial in the case of erasures, the syndrome is replaced with the modified syndromes S' in the equation for Δ_r:

$$\Delta_r = \sum_{j=0}^{n-1} \Lambda_j^{(r-1)} S'_{r-1-j}.$$

After n iterations starting with the initial values $\Lambda^{(0)}(x) = B^{(0)}(x) = 1$, the error locator polynomial $\Lambda(x)$ is obtained.

However, it is much better to bury the erasure polynomial in the initialization of the Berlekamp–Massey algorithm. Start instead with the values $\Lambda^{(0)}(x) = B^{(0)}(x) = \Psi(x)$. Then the iterations compute

$$\left[\begin{array}{c} \Lambda^{(r)}(x)\Psi(x) \\ B^{(r)}(x)\Psi(x) \end{array} \right] = \left[\begin{array}{cc} 1 & -\Delta_r x \\ \delta_r \Delta_r^{-1} & (1 - \delta_r)x \end{array} \right] \left[\begin{array}{c} \Lambda^{(r-1)}(x)\Psi(x) \\ B^{(r-1)}(x)\Psi(x) \end{array} \right].$$

Temporarily define $\overline{\Lambda}^{(r)}(x) = \Lambda^{(r)}(x)\Psi(x)$ and compute Δ_r by

$$\Delta_r = \sum_{j=0}^{n-1} \overline{\Lambda}_j^{(r-1)} E_{r-1-j} = \sum_{j=0}^{n-1} \overline{\Lambda}_{r-1-j}^{(r-1)} E_j.$$

This is the correct value of Δ_r because

$$\Delta_r = \sum_{j=0}^{n-1} \left(\sum_{k=0}^{n-1} \Lambda_k^{(r-1)} \Psi_{r-1-j-k} \right) E_j = \sum_{k=0}^{n-1} \Lambda_k^{(r-1)} S'_{r-1-k}.$$

Therefore if we initialize the Berlekamp–Massey algorithm with $\Psi(x)$, the modified syndromes are implicit and need not explicitly appear. The notation $\overline{\Lambda}(x)$ can now be replaced with $\Lambda(x)$, which will now be called the error and erasure locator polynomial. The Berlekamp–Massey algorithm initialized with $\Psi(x)$ generates recursively the error and erasure locator polynomial using exactly the same equations as for the case of errors-only decoding. The only change from the decoder for errors only is the computation of the erasure locator polynomial, which is trivial compared to other decoding computations.

8.3 Time Domain Decoder Algorithms

The decoding of Reed–Solomon codes has been developed as a computation in the Fourier transform domain, but there are other possibilities for the processing. In this section we shall examine the image of the Berlekamp–Massey algorithm in the time domain.

Because it processes the spectrum of the received word, the Berlekamp–Massey algorithm is preceded by a Fourier transform and is followed by some form of an inverse Fourier transform. However, instead of pushing the received word into the frequency domain for processing, it is possible to push the Berlekamp–Massey algorithm into the time domain. This makes the Fourier transforms simply vanish from the computation. On the other hand, the frequency domain vectors of length t are replaced by time domain vectors of length n. Algorithms in the frequency domain may have $2t$ or n iterations of processing vectors of length t; we say they have complexity $2t^2$

or nt. Algorithms in the time domain have $2t$ or n iterations of processing vectors of length n; we say they have complexity $2tn$ or n^2. The time domain decoders are structurally simple and are useful in applications where structural simplicity is important and the number of iterations is not.

8.3.1   A TIME DOMAIN DECODER WITH n^2 STEPS

Let the vectors $\boldsymbol{\lambda}$ and $\mathbf{b}$ denote the inverse Fourier transform of the vectors Λ and $\mathbf{B}$, respectively. To push the Berlekamp–Massey equations into the time domain, simply replace the frequency domain variables Λ_k and B_k with the time domain variables λ_i and b_i, replace the delay operator x with multiplication by ω^{-i}, and replace componentwise products with convolutions. Replacement of the delay operator with multiplication by ω^{-i} is justified by the translation property of the Fourier transform; replacement of a componentwise product with a convolution is justified by the convolution theorem. Then the time domain algorithm is as follows.

Let $\mathbf{v}$ be the received noisy Reed–Solomon codeword and let the following set of recursive equations be used to compute $\lambda_i^{(2t)}$ for $i = 0, \ldots, n - 1$:

$$\Delta_r = \sum_{i=0}^{n-1} \omega^{i(r-1)} [\lambda_i^{(r-1)} v_i],$$

$$L_r = \delta_r (r - L_{r-1}) + (1 - \delta_r) L_{r-1},$$

$$\begin{bmatrix} \lambda_i^{(r)} \\ b_i^{(r)} \end{bmatrix} = \begin{bmatrix} 1 & -\Delta_r \omega^{-i} \\ \delta_r \Delta_r^{-1} & (1 - \delta_r) \omega^{-i} \end{bmatrix} \begin{bmatrix} \lambda_i^{(r-1)} \\ b_i^{(r-1)} \end{bmatrix}$$

for $i = 0, \ldots, n - 1$ and $r = 1, \ldots, 2t$. The initial conditions are $\lambda_i^{(0)} = 1$ for all i, $b_i^{(0)} = 1$ for all i, $L_0 = 0$, and $\delta_r = 1$ if both $\Delta_r \neq 0$ and $2L_{r-1} \leq r - 1$, and otherwise $\delta_r = 0$. Then $\lambda_i^{(2t)} = 0$ if and only if $e_i \neq 0$.

For nonbinary codes it is not enough to compute only the error locations; we must also compute the error magnitudes. In the frequency domain these are computed by the following recursion:

$$E_k = -\sum_{j=1}^{t} \Lambda_j E_{k-j}, \qquad k = 2t, \ldots, n - 1.$$

It is not possible just to write the Fourier transform of this equation; some restructuring is necessary. The following equivalent set of recursive equations for $r = 2t, \ldots, n - 1$ is suitably restructured:

$$\Delta_r = \sum_{i=0}^{n-1} \omega^{ir} v_i^{(r-1)} \lambda_i,$$

$$v_i^{(r)} = v_i^{(r-1)} - \Delta_r \omega^{-ri}.$$

Starting with $v_i^{(2t)} = v_i$, and $\lambda_i = \lambda_i^{(2t)}$ for $i = 0, \ldots, n-1$, the last iteration results in

$$v_i^{(n)} = e_i, \qquad i = 0, \ldots, n - 1.$$

This works because $E_k = V_k$ for $k = 0, \ldots, 2t - 1$, and the new equations, although written in the time domain, are in effect changing sequentially V_k to E_k for $k = 2t, \ldots, n - 1$.

A time domain decoder with n^2 steps is shown in Figure 8.2. Provision for decoding a singly extended Reed–Solomon code is included by a special test at iteration $2t - 2$. Provision for decoding erasures as well as errors is included by providing a special loop for the first ρ iterations. The index r is used to count out the first ρ iterations while the erasure polynomial is being formed and then continues to count out the iterations of the Berlekamp–Massey algorithm, stopping when r reaches $2t$. The shift register length L is increased once for each erasure, and thereafter the length changes according to the procedure of the Berlekamp–Massey algorithm. Because this algorithm was developed in the absence of erasures, however, the length test

$$L \longleftarrow r - L \quad \text{if } 2L \leq r - 1$$

must be modified by replacing r and L by $r - \rho$ and $L - \rho$, respectively.

The time domain decoder with complexity n^2 has no Fourier transforms (no syndrome computation or Forney algorithm); it has only one major computational block which is easily designed into digital logic. It does, however, always deal with vectors of length n rather than with the vectors of length t used by the frequency domain decoder. Hence, there are hardware/speed tradeoffs.

8.3.2   A TIME DOMAIN DECODER WITH $2tn$ STEPS

To get a faster decoder in the time domain, we can start with the Berlekamp algorithm. Let $\boldsymbol{\lambda}'$ and $\mathbf{b}'$ be the inverse Fourier transforms of Λ' and $\mathbf{B}'$ corresponding to $\Lambda'(x)$ and $B'(x)$, the formal derivatives of $\Lambda(x)$ and $B(x)$. The vectors $\boldsymbol{\lambda}'$ and $\mathbf{b}'$ will be computed iteratively in the time domain to avoid the need to enter the frequency domain to compute the formal derivative. Transformed into the time domain, the full set of equations for the iteration becomes the following:

$$
\begin{bmatrix} \lambda_i^{(r)} \\ b_i^{(r)} \\ \lambda_i'^{(r)} \\ b_i'^{(r)} \end{bmatrix}
=
\begin{bmatrix}
1 & -\Delta_r \omega^{-i} & 0 & 0 \\
\delta_r \Delta_r^{-1} & (1 - \delta_r)\omega^{-i} & 0 & 0 \\
0 & -\Delta_r & 1 & -\Delta_r \omega^{-i} \\
0 & (1 - \delta_r) & \delta_r \Delta_r^{-1} & (1 - \delta_r)\omega^{-i}
\end{bmatrix}
\begin{bmatrix} \lambda_i^{(r-1)} \\ b_i^{(r-1)} \\ \lambda_i'^{(r-1)} \\ b_i'^{(r-1)} \end{bmatrix},
$$

$$
\begin{bmatrix} \gamma_i^{(r)} \\ a_i^{(r)} \end{bmatrix}
=
\begin{bmatrix}
1 & -\Delta_r \omega^{-i} \\
\delta_r \Delta_r^{-1} & (1 - \delta_r)\omega^{-i}
\end{bmatrix}
\begin{bmatrix} \gamma_i^{(r-1)} \\ a_i^{(r-1)} \end{bmatrix},
$$

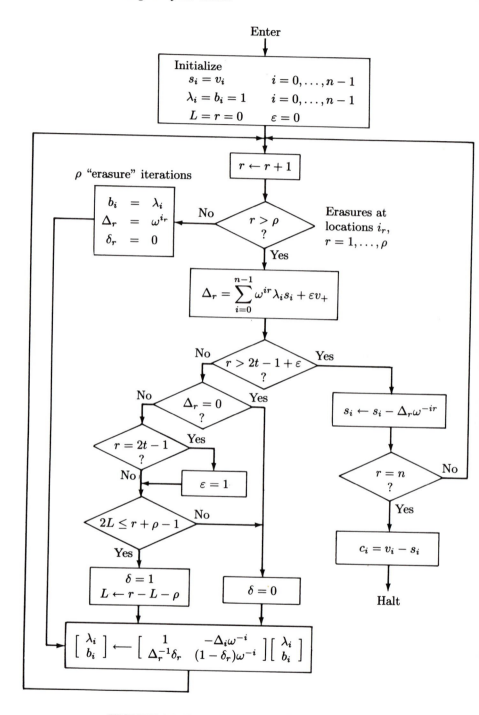

FIGURE 8.2. Time domain decoder algorithm.

for $i = 0, \ldots, n - 1$, and $r = 1, \ldots, 2t$. The initial conditions are $\lambda_i^{(0)} = b_i^{(0)} = \gamma_i^{(0)} = 1$ for all i; $\lambda_i'^{(0)} = b_i'^{(0)} = a_i^{(0)} = 0$ for all i; $L_0 = 0$; and $\delta_r = 1$ if both $\Delta_r \neq 0$ and $2L_{r-1} \leq r - 1$, and otherwise $\delta_r = 0$.

After $2t$ iterations, the zeros of $\boldsymbol{\lambda}$ point to the error locations and the ratio γ_i/λ_i' gives the magnitude of the error whenever $\lambda_i = 0$.

8.3.3 A TIME DOMAIN DECODER WITH (MORE THAN) $4t^2$ STEPS

If the blocklength is composite, $n = n'n''$, and n' is not smaller than $2t$, it is possible to "fold up" the received vector into a vector of length n' which is then corrected in about $(n')^2$ or $2tn'$ steps using either procedure already described. The folded error vector of length n' then must be unfolded (interpolated) to obtain the desired vector $\mathbf{e}$ of blocklength n. The final step of interpolation does have $n'n$ steps so the complete decoder has about $2tn$ steps. However, interpolation is simpler than the Berlekamp–Massey algorithm, so there can be a net improvement.

To see how to fold the data vector, recall that the decoder solves for $\Lambda(x)$ and possibly $\Lambda'(x)$ and $\Gamma(x)$ given only the syndromes $E_0, \ldots, E_{2t-1}$. The polynomial $\Lambda(x)$ satisfies

$$\sum_{j=0}^{t} \Lambda_j E_{k-j} = 0, \qquad k = t, \ldots, 2t - 1$$

and the polynomial $\Gamma(x)$ satisfies

$$\sum_{j=0}^{t} \Lambda_j E_{k-j} = \Gamma_k, \qquad k = 0, \ldots, t - 1.$$

While the syndrome vector is actually computed as $2t$ components of a Fourier transform of blocklength n, it would make no difference in the computation of $\Lambda(x)$ or of $\Gamma(x)$ if instead it came from a Fourier transform of blocklength n' provided n' is not smaller than $2t$.

Imagine for the moment the following procedure. Given the received vector $\mathbf{v}$ (of blocklength n) compute the Fourier transform of blocklength n. Take the segment of length $2t$ including the syndromes and take the inverse Fourier transform of blocklength n'. This gives a new vector $\mathbf{v}'$ of blocklength n' that has the property that if $\mathbf{v}'$ is used as the input to a Reed–Solomon decoder for codes of blocklength n', the same $\Lambda(x)$ and $\Gamma(x)$ will be computed.

A time domain decoder can be used for $\mathbf{v}'$ producing the vectors $\boldsymbol{\lambda}_0$, $\boldsymbol{\lambda}_0'$, and $\boldsymbol{\gamma}_0$, all of blocklength n'. These can be computed with $2tn'$ iterations of a time domain Berlekamp algorithm. An n'-point Fourier transform produces $\Lambda(x)$, $\Lambda'(x)$, and $\Gamma(x)$. An n-point inverse Fourier transform then produces $\boldsymbol{\lambda}$, $\boldsymbol{\lambda}'$, and $\boldsymbol{\gamma}$, all of blocklength n, from which the error pattern is computed.

Finally, we recall the FFT algorithms for subblocks discussed in Section 7.4. Computation of $\mathbf{v}'$ from $\mathbf{v}$ has the form

$$\mathbf{v}' = [\mathbf{I} \quad \mathbf{A} \quad \mathbf{A}^2 \ \ldots \ \mathbf{A}^{n''-1}]\mathbf{v},$$

where $\mathbf{A}$ is an $n' \times n'$ circulant matrix. The matrix-vector product can be parenthesized by blocks using Horner's rule and each block matrix-vector product has the form of a cyclic convolution. Thus $\mathbf{v}'$ can be computed with $n'' - 2$ cyclic convolutions of blocklength n'.

The recovery of $\boldsymbol{\lambda}$ (or $\boldsymbol{\lambda}'$, or $\boldsymbol{\gamma}$) from $\boldsymbol{\lambda}_0$ (or $\boldsymbol{\lambda}_0'$ or $\boldsymbol{\gamma}_0'$) has the form of an interpolation and can be computed as discussed in Section 7.4 using $n'' - 2$ cyclic convolutions each of blocklength n'.

The computations can be particularly graphic for a binary code. A received noisy codeword of blocklength n over $GF(2)$ is folded into an equivalent noisy codeword of blocklength n' over the extension field $GF(2^m)$. This field is where most of the decoding action takes place.

8.4   A Universal Decoder Architecture

The usefulness of the signal processing view of error-control codes and the power of the Fourier transform are underscored by describing a universal decoder architecture. This architecture also serves to recapitulate the decoder algorithms discussed in this chapter. In some sense it forms a capstone for the book because it gives a decoder with the spirit of digital signal processing.

The architecture is based on the flow diagram shown in Figure 8.3. This flow diagram gives a time domain version of the Berlekamp algorithm, but also including erasure decoding and the decoding of a single extension symbol. Most of the clutter in Figure 8.3 is concerned with logical tests and the setting of switches, and it is quite trivial in a hardware implementation of a decoder.

The flow diagram is best understood by following the r index, which counts out the $2t$ main iterations. During the first ρ iterations, with ρ equal to the number of erased symbols, the basic iteration is tricked into initializing itself for ρ erasures, as was described in Section 8.2. This is done with the same computations as would be done if there were no erasures, except that different variables are switched into the input of the computations. There is virtually no increase in complexity to fill erasures. After $2t$ iterations, the time domain iterates are computed. It only remains to compute the error vector by Forney's algorithm as shown on the right side.

The structure of the flow diagram in Figure 8.3 does not depend on t; the only effect of changing t is that more iterations of the same kind are

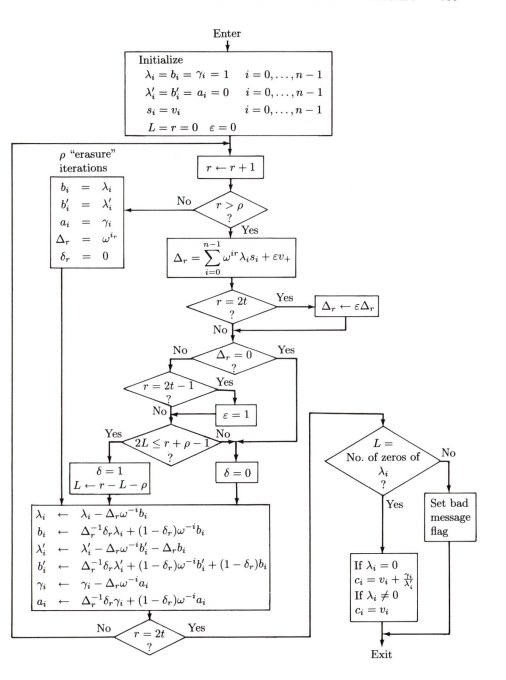

FIGURE 8.3. A time domain version of the Berlekamp algorithm.

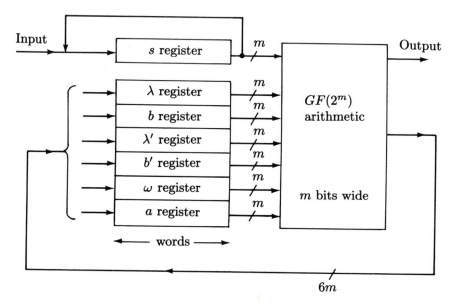

FIGURE 8.4. The architecture of a universal Reed–Solomon decoder.

performed. Consequently this flow diagram is well suited to a universal decoder, one that can decode any Reed–Solomon (or BCH) code within some maximum limits. Shortened codes are also readily decoded.

Figure 8.4 shows the architecture of a universal Reed–Solomon decoder. With 256 chosen as the maximum blocklength and $GF(256)$ as the maximum binary symbol field, the decoder consists of seven registers — eight bits wide and 255 symbols long (and one extension symbol for the received word). These registers are thought of as shift registers but can be implemented as random-access memories. The registers hold the $\mathbf{v}$, $\boldsymbol{\lambda}$, $\mathbf{b}$, $\boldsymbol{\lambda}'$, $\mathbf{b}'$, $\boldsymbol{\gamma}$, and $\mathbf{a}$ vectors.

There are seven symbols (56 bits) moving in parallel through the decoder and circulating back to the shift registers. Within the decoder are a fixed set of Galois field multipliers and adders, forming Δ and updating the various stored vectors in accordance with Figure 8.3. The multipliers may be designed so that any binary field up to $GF(256)$ can be selected for the application.

References

[1] S. Arimoto, Encoding and decoding of p-ary group codes and the correction system (in Japanese), *Inform. Process. in Japan*, 2:321–325 (1961).

[2] R. C. Agarwal and C. S. Burrus, Fast convolution using Fermat number transforms with applications to digital filtering, *IEEE Trans. Acoust. Speech Signal Process.*, 22:87–97 (1974).

[3] _____, Number theoretic transforms to implement fast digital convolution, *Proc. IEEE*, 63:550–560 (1975).

[4] E. R. Berlekamp, *Algebraic Coding Theory*. McGraw-Hill, New York, 1968.

[5] G. Birkhoff and S. MacLane, *A Survey of Modern Algebra*, revised ed., Macmillan, New York, 1953.

[6] R. E. Blahut, Transform techniques for error control codes, *IBM J. Res. Devel.*, 23:299–315 (1979).

[7] R. E. Blahut, *Theory and Practice of Error Control Codes*, Addison-Wesley, Reading, MA, 1983.

[8] R. E. Blahut, Fast convolution of rational sequences, in *Abstracts — 1983 IEEE Int. Symp. Inform. Theory*, St. Jovite, Quebec, Canada, 1983.

[9] R. E. Blahut, A Universal Reed–Solomon Decoder, *IBM J. Res. Devel.*, 28:150–158 (1984).

[10] R. E. Blahut, *Fast Algorithms for Digital Signal Processing*, Addison-Wesley, Reading, MA, 1985.

[11] R. E. Blahut, Algebraic fields, signal processing, and error control, *Proc. IEEE*, 73:874–893 (1985).

[12] R. E. Blahut, The Gilbert–Prange Theorem, *IEEE Trans. Inform. Theory*, 37:1269–1273 (1991).

[13] P. R. Chevillat, Transform-domain filtering with number theoretic transforms, and limited word length, *IEEE Trans. Acoust. Speech Signal Process.*, 26:284–290 (1978).

[14] T. K. Citron, *Algorithms and Architectures for Error-Correcting Codes*, Ph.D Dissertation, Stanford University, Stanford, CA, 1986.

[15] J. W. Cooley and J. W. Tukey, An algorithm for the machine computation of complex Fourier series, *Math. Comp.*, 19:297–301 (1965).

[16] J. H. Cozzens and L. A. Finkelstein, Computing the discrete Fourier transform using residue number systems in a ring of algebraic integers, *IEEE Trans. Inform. Theory*, 31:580–588 (1985).

[17] J. Durbin, The fitting of time-series models, *Rev. Internat. Statist. Inst.*, 23:233–244 (1960).

[18] W. L. Eastman, Inside Euclid's Algorithm, in *Coding Theory and Design Theory, Part I: Coding Theory*, Dijen Ray-Chaudhuri, ed., Springer-Verlag, New York, 1990.

[19] R. A. Games, Complex approximations using algebraic integers, *IEEE Trans. Inform. Theory*, 31:565–579 (1985).

[20] I. J. Good, The interaction algorithm and practical Fourier analysis, *J. Roy. Statist. Soc. Ser. B*, 20:361–375 (1958); addendum, 22:372–375 (1960).

[21] C. R. P. Hartmann and K. K. Tzeng, Generalizations of the BCH bound, *Inform. and Control*, 20:489–498 (1972).

[22] I. N. Herstein, *Topics in Algebra*, Balisdell, Waltham, MA, 1964.

[23] D. P. Kolba and T. W. Parks, A prime factor FFT algorithm using high speed convolution, *IEEE Trans. Acoust. Speech Signal Process.*, 25:281–284 (1977).

[24] A. Lempel and S. Winograd, A new approach to error-correcting codes, *IEEE Trans. Inform. Theory*, 23:503–508 (1977).

[25] N. Levinson, The Wiener RMS error criterion in filter design and prediction, *J. Math. Phys.*, 25:261–278 (1947).

[26] M. W. Marcellin and T. R. Fischer, Encoding algorithms for complex approximations in $Z[e^{2\pi i/8}]$, *IEEE Trans. Inform. Theory*, 35:1133–1136 (1989).

[27] T. G. Marshall, Jr., Coding of real-number sequences for error correction: A digital signal processing problem, *IEEE J. Selected Areas Comm.*, 2:381–392 (1984).

[28] J. L. Massey, Shift-register synthesis and BCH decoding, *IEEE Trans. Inform. Theory*, 15:122–127 (1969).

[29] U. Maurer and R. Viscardi, *Running-Key Generators with Memory in the Nonlinear Combining Function*, Diploma thesis, Swiss Federal Institute of Technology, Zurich, 1984.

[30] J. H. McClellan and C. M. Rader, *Number Theory in Digital Signal Processing*, Prentice-Hall, Englewood Cliffs, NJ, 1979.

[31] H. Niederreiter, A simple and general approach to the decimation of feedback shift-register sequences, *Problems Control Inform. Theory*, 17:327–331 (1988).

[32] H. J. Nussbaumer, Digital filtering using complex Mersenne transforms, *IBM J. Res. Devel.*, 20:498–504 (1976).

[33] _____, Digital filtering using polynomial transforms, *Electron. Lett.*, 13:386–387 (1977).

[34] _____, *Fast Fourier Transform and Convolutional Algorithms*, 2nd ed., Springer-Verlag, Berlin, 1982.

[35] A. V. Oppenheim and R. W. Shafer, *Digital Signal Processing*, Prentice-Hall, Englewood Cliffs, NJ, 1975.

[36] I. M. Pollard, The fast Fourier transform in a finite field, *Math. Comp.*, 25:365–374 (1971).

[37] L. R. Rabiner and B. Gold, *Theory and Application of Digital Signal Processing*, Prentice-Hall, Englewood Cliffs, NJ, 1975.

[38] C. M. Rader, Discrete Fourier transforms when the number of data samples is prime, *Proc. IEEE*, 56:1107–1108 (1968).

[39] C. M. Rader, Discrete convolutions via Mersenne transforms, *IEEE Trans. Comput.*, 21:1269–1273 (1972).

[40] I. S. Reed and G. Solomon, Polynomial codes over certain finite fields, *J. Soc. Indust. Appl. Math.*, 8:300–304 (1960).

[41] I. S. Reed and T. K. Truong, The use of finite fields to compute convolutions, *IEEE Trans. Inform. Theory*, 21:208–213 (1975).

[42] B. Rice, Winograd convolution algorithms over finite fields, *Congressus Numberatium*, 29:827–857 (1980).

[43] C. Roos, A new lower bound for the minimum distance of a cyclic code, *IEEE Trans. Inform. Theory*, 29:330–332 (1983).

[44] R. M. Roth and A. Lempel, Application of circulant matrices to the construction and decoding of linear codes, *IEEE Trans. Inform. Theory*, 36:1157–1163 (1990).

[45] D. V. Sarwate, and M. B. Pursley, Crosscorrelation properties of pseudorandom and related sequences, *Proc. IEEE*, 68:593–619 (1980).

[46] T. Schaub, *A Linear Complexity Approach to Cyclic Codes*, Doctor of Technical Sciences Dissertation, ETH Swiss Federal Institute of Technology, 1988.

[47] B. Smeets, *Some Results on Linear Recurring Sequences*, Ph.D. dissertation, University of Lund, 1987.

[48] Y. Sugiyama, M. Kasahara, S. Hirasawa, and T. Namekawa, A method for solving key equations for decoding Goppa codes, *Inform. Control*, 27:87–99 (1975).

[49] L. H. Thomas, Using a computer to solve problems in physics, in *Applications of Digital Computers*, Ginn and Co., Boston, MA, 1963.

[50] W. F. Trench, An algorithm for the inversion of finite Toeplitz matrices, *J. Soc. Indust. Appl. Math.*, 12:512–522 (1964).

[51] B. L. van der Waerden, *Modern Algebra* (2 vols), transl. by F. Blum and T. J. Benac, Frederick Unger, New York, 1950 and 1953.

[52] J. H. van Lint, *Introduction to Coding Theory*, Springer-Verlag, Berlin, 1982.

[53] J. H. van Lint and R. M. Wilson, On the minimum distance of cyclic codes, *IEEE Trans. Inform. Theory*, 32:23–40 (1986).

[54] S. Winograd, On computing the discrete Fourier transform, *Proc. Nat. Acad. Sci. USA*, 73:1005–1006 (1976).

[55] ———, *Arithmetic complexity of computations*, volume 33 of *CBMS-NSF Regional Conf. Ser. Appl. Math.*, Society for Industrial and Applied Mathematics, Philadelphia, PA, 1980.

[56] J. K. Wolf, Adding two information symbols to certain nonbinary BCH codes and some applications, *Bell Systems Tech. J.*, 48:2405–2424 (1969).

[57] ———, Redundancy, the discrete Fourier transform, and impulse noise cancellation, *IEEE Trans. Comm.*, 31:458–461 (1983).

[58] N. Zierler, Linear recurring sequences, *J. Soc. Indust. Appl. Math.*, 7:31–48 (1959).

Index